S0-BLA-805

CHEMICAL ANALYSIS OF SILICATE ROCKS

SERIES

Methods in Geochemistry and Geophysics

1. A.S. RITCHIE
CHROMATOGRAPHY IN GEOLOGY

2. R. BOWEN
PALEOTEMPERATURE ANALYSIS

3. D.S. PARASNIS
MINING GEOPHYSICS

4. I. ADLER
X-RAY EMISSION SPECTROGRAPHY IN GEOLOGY

5. THE LORD ENERGLYN AND L. BREALEY
ANALYTICAL GEOCHEMISTRY

7. E.E. ANGINO AND G.K. BILLINGS
ATOMIC ABSORPTION SPECTROMETRY IN GEOLOGY

8. A. VOLBORTH
ELEMENTAL ANALYSIS IN GEOCHEMISTRY
A. MAJOR ELEMENTS

9. P.K. BHATTACHARYA AND H.P. PATRA
DIRECT CURRENT GEOELECTRIC SOUNDING

10. J.A.S. ADAMS AND P. GASPARINI
GAMMA-RAY SPECTROMETRY OF ROCKS

11. W. ERNST
GEOCHEMICAL FACIES ANALYSIS

Methods in Geochemistry and Geophysics

6

CHEMICAL ANALYSIS OF SILICATE ROCKS

BY

A.J. EASTON

British Museum (Natural History), London, Great Britain

ELSEVIER PUBLISHING COMPANY
AMSTERDAM / LONDON / NEW YORK
1972

ELSEVIER PUBLISHING COMPANY
335 Jan van Galenstraat
P.O. Box 211, Amsterdam, The Netherlands

AMERICAN ELSEVIER PUBLISHING COMPANY, INC.
52 Vanderbilt Avenue
New York, New York 10017

Library of Congress Card Number: 67-25751
ISBN 0-444-40985-8
With 24 illustrations, 4 schemes and 13 tables

Printed in The Netherlands

Preface

Since the scope of geological investigation has expanded to include geochemistry it has been necessary to increase the number of analyses for the major elements of all types of silicate rocks to supply the required data. Often a study is required not only of the composition of the total rock but also of the distribution of the major elements between the various phases of which the rock is composed. Such studies are now being used to probe into the problems of the earth's crust and upper mantle.

A number of M.Sc. and Ph.D. theses in geophysics now include a study of the chemical relationship between geological samples. Students therefore find it essential to be able to carry out chemical analyses of the principal types of silicate rocks and minerals themselves. However, to them who have taken only a preliminary course in general chemistry it will be difficult, in a limited period of study, to sift and evaluate the large number of analytical papers being published on this subject. There is no lack of methods that are satisfactory when applied to the analysis of pure compounds; there is more need for quantitative procedures that can be applied to the separation and determination of elements in complex mixtures as in silicate rocks. Therefore, I have endeavoured to present the procedures that appear to have the most merit in silicate material and tried to discuss each procedure at sufficient length, for help to students interested in silicate analysis. Further, frequent references are introduced to important papers on the different subjects discussed, in order to direct the student's attention to the original literature. The inclusion of methods and schemes suitable for the examination of meteoric silicate material may be of assistance to those who are pursuing investigation in this section of geochemistry.

A.J. EASTON

ACKNOWLEDGEMENT

The text was compiled during the time the author was resident in Australia. At this time he was engaged on analysis of silicate materials for various members of the Department of Geophysics, Australian National University, Canberra and he wishes to express his gratitude to all those who encouraged the project. In particular he wishes to thank Dr. L.P. Greenland, Dr. K.H. Wolf and Dr. M.J. Abbott who at various stages of the manuscript kindly offered constructive criticism, and also the late Dr. J.W. Rees who was closely associated with the text in its final form.

Contents

CHAPTER 1

Introduction

PRELIMINARY REMARKS

For many years the order of treatments according to the classical scheme of Washington (1919) using gravimetric and volumetric procedures only has been followed, since they are generally very accurate when large quantities are involved. During the past decennia in the progress of chemistry, besides the micro balance with a high degree of sensitivity, new techniques (colorimetric and polarographic) and more selective reagents for detection, quantitative separation and estimation, have been gradually developed or discovered. These wet techniques (visual colorimetry, visual nephelometry, absorptiometry, spectrophotometry, and electrolysis with cathode of falling mercury) in turn have been used in analyses where speed, convenience and relatively high precision and accuracy in the determination of micro and semimicro quantities of constituents are of consequence. This development has tended to split silicate analysts into two groups, each vigorously supporting either the general scheme used by Washington or "rapid methods" schemes involving a minimum number of operations. In actual fact sound modifications of the Washington's procedures and judicious applications of supplementary "rapid methods" on a quantitative micro- or semimicrochemical basis appear to offer the best combination for usefulness and accuracy.

Spectrophotometric methods are now established inter alii for the determination of chromium, cobalt, copper, iron (in the ferrous and ferric states), manganese, nickel, phosphorus, silicon, and titanium. Recent developments in this field have been the elimination of interferences from the determinations, e.g., the retaining of iron (one of the most common interfering elements) from the solution under test by a simple procedure on an ion-exchange resin as a chloride complex offers a tremendous advantage over the older methods of separation. The introduction of EDTA (ethylenediamine tetraacetic acid) as either a complexing or reducing agent has allowed the determination of small quantities of chromium and manganese in the presence of a large amount of iron. The use of EDTA as a titrant has permitted an alternative to the gravimetric method for the determination of calcium and magnesium. As with most new analytical methods a wide variety of organic dye indicators has been proposed as most suitable for the indication of the end point of the titration. Some authors have proposed separation of calcium before titration of magnesium, others prefer a titration of the combined calcium and magnesium, subtracting the calcium which has been determined on a separate titration using specific conditions and indicator. The EDTA-titration is used to advantage when only small quantities of these two elements are present which, if determined gravimetrically, would be liable to a high degree of contamination by salts.

For accurate work the gravimetric method is still used for the determination of silicon, the most common element. The unrecovered silicon which passes through the final filtration is determined spectrophotometrically. The entirely spectrophotometric method for silicon finds an ideal application where a suite of rocks of similar composition is required to be analyzed and the silicon has been determined gravimetrically as silica on typical specimens. These are then used as standard materials for the spectrophotometric determination of the other samples. In mineral analyses it is most important to obtain a correct value for silicon as this will largely control the quantity of the other elements to be placed in the same grouping as silicon for the purpose of calculating atomic formulae.

The determination of aluminium has alway posed a problem where it is a fairly minor constituent of the ammonium hydroxide group. Where aluminium is determined by "difference", errors in the determination of the other constituents (particularly where these have not been determined on the ignited ammonium hydroxide precipitate) tend to make this the worst determined element in silicate material. This contention is supported by the study on results reported by various analysts for the W1 and G1 rocks (Stevens and Niles, 1960). It is suggested here that the iron is taken out by ion-exchange resin or ether extraction and that the other constituents of the ammonium hydroxide precipitate are determined on a subsequent potassium bisulphate fusion of the precipitate (except phosphorus).

Flamephotometric determination of the alkali and alkaline earth elements has now replaced the previous gravimetric methods, the main advantage having been gained in the determination of the lower concentrations. The instrumental method has not been without its problems; different design burners and use of different combination of gases for the excitation of the elements have made a single procedure difficult to formulate. An accurate determination of potassium is of particular importance where the result is required for K/Ar age determination. A large number of factors have been successfully investigated for the purpose of obtaining increased accuracy (Cooper, 1963) and anyone attempting potassium determinations for age work would be well advised to consult the original study.

Determination of water in rocks substantially free from volatile constituents other than water or carbon dioxide, by absorption of the liberated vapour on filter paper ("modified Penfield method", Harvey, 1939) has been modified so that the sample is heated at a controlled temperature and the water vapour absorbed into weighed tubes containing water absorbents. This modification has also allowed the liberated carbon dioxide to be collected in a weighed absorbent on the same sample as the water. These two determinations have replaced the old-time "loss on ignition" method for Total water which is affected in most cases by serious errors.

Perhaps one of the most challenging problems in silicate work is the analysis of meteoritic silicate material, e.g., of chondrites. H.B.Wiik (Finnish Geological Survey) has laid a firm foundation for the analysis of this material using modern methods. Interest in this extraterrestrial material has recently increased due to the postulation of the chondritic earth model. Analysis of meteoritic iron also has an important place in the study of extraterrestrial materials.

In addition to being of use in geological interpretation, modern chemical methods may be used with advantage to establish the composition of standard materials being of great help in checking i.a. physical methods such as those based on absorption or emission spectra or X-ray phenomena. In these cases the accuracy of the results obtained with the material under test is dependent on the value of the analysis made on the standard material.

Testing materials it must be emphasized that an accurate analysis will be only possible if the sample truly represents the material under test. Therefore the selection of correct portions of the material and the preparation of a representative test sample from such portions are necessary prerequisites to every analysis. Especially when small samples are taken for analysis the test sample must be perfectly homogeneous to represent the true composition of the material from which it was selected.

CHAPTER 2

Apparatus and Reagents

BALANCE AND WEIGHTS

The evolution of chemical analysis is very much a story of the gradual development of the chemical balance because the value of a chemical analysis rests fundamentally on the accuracy of the balance and the weights that are used, culminating of late decennia in the micro balance with a sensitivity of a thousandth of a milligram. For details concerning the mounting and lightning of the balance, its sensitivity, the choice of a good set of weights, testing and care of balance and weights, and trustworthy methods of weighing, see textbooks of quantitative analysis, e.g., Treadwell and Hall (1935), and Kolthoff and Sandell (1947).

GLASSWARE AND PORCELAIN

All glassware and porcelain used in analytical work should be carefully selected to meet the particular requirements for each operation. Well-known brands of chemical glassware are Pyrex, Jena, Duran, and Kimble; they differ in composition and in properties. So-called resistant glass is attacked by all solutions; its resistance to alkaline solutions or to pure water is less than to acid solutions. In general good porcelain is more resistant to solutions than glass. For this reason, alkaline or neutral solutions kept in glass should be acidified whenever possible or should be kept in vessels of good porcelain when they must stand over any length of time. Owing to the fact that the compositions of glass and porcelain are quite different, the amount and nature of substances extracted from the vessels and contaminating the material under test, may vary.

Errors caused by substances extracted from glassware or porcelain should be corrected by running blanks undergoing the same reactions as the material under test, or by checking the results of the analysis against those of comparable samples of known composition.

Fritted-glass filtering crucibles, funnels, cover glasses, glass rods, etc., of chemical-resistant glass should be used.

Dark-coloured glass for the protection of solutions affected by light, alkali-resistant glass for use where superior resistance to alkalies is important, and high-silica glass having exceptional resistance to thermal shocks are available.

Standard volumetric flasks, burettes, and pipettes should be of precision grade. For general specifications, methods of calibration, and testing of glass volumetric apparatus, see e.g., Stott (1928), or Kolthoff and Stenger (1947).

PLATINUM AND SUBSTITUTES

Platinum ware

Because of its high melting point (1,770°C) and its resistance to chemicals, platinum is indispensable in any chemical laboratory. Platinum metal is required or recommended for: crucibles for high-temperature ignitions (to about 1,200°C) and for special fusions (e.g., with Na_2CO_3 or $K_2S_2O_7$), dishes for dissolution of alkaline or acidic smelts and for evaporating of solutions, cones and dishes for Gooch crucibles, electrodes for electrodeposition methods using platinum electrodes, triangles supporting platinum crucibles during ignitions, platinum-tipped crucible tongs or forceps to bring hot platinum vessels into the desired position, rods for stirring viscous melts or solutions containing hydrofluoric acid, wire for making bead or flame tests, plating of weights of other metals that tarnish on exposure to the atmosphere, material for the small weights as low as 1 mg, etc.

Because of the common use of platinum in any analytical work, the following practical points must be regarded:

(*1*) Platinum ware should never be heated in contact with metals other than platinum; other metals will alloy with heated platinum to give an alloy of much lower melting point thus damaging or even destroying the ware. Platinum-tipped tongs or forceps should be used at all times when handling heated platinum; when not in use the tongs should be placed with the platinum tips uppermost to avoid contamination.

(*2*) Direct contact of heated platinum with unburned gas or sooty flames leads to the formation of a carbide of platinum rendering the metal brittle; using gas burners heating should always be done above the blue or green zone of the nonluminous flame from burners of the Bunsen- or Meker type.

(*3*) Products of reduction (antimony, arsenic, phosphorus, lead, etc.) and compounds of easily reducible metals which combine with the platinum vessels in which the ignitions or fusions are made must be avoided, as must substances such as sulphides, phosphides, or arsenides.

(*4*) Caustic alkalies and barium oxide attack platinum strongly, likewise their hydroxides, nitrates, nitrites, and cyanides.

(*5*) Mixtures of nitrates and chlorides, or chlorides and nitric acid, yielding aqua regia must be avoided, as must other combinations such as hydrochloric acid and e.g., highly ferruginous substances; manganese ores in more or less peroxidized conditions; certain basic oxides of the rare earth elements; compounds of manganese, chromium and vanadium oxidized by fusions to manganate, chromate, and vanadate; all liberating chlorine. The same holds for combinations yielding bromine.

(*6*) Alkali carbonates fusions attack platinum slightly when heated over a free flame, and to a somewhat greater extent when heated in a muffle.

(*7*) With pyrosulphate fusions the attack is more marked; when necessary, the dissolved platinum in an ordinary fusion (averaging about 3 mg) or resulting from other treatments in platinum ware, may be removed at some subsequent stage of analysis by hydrogen sulphide from hydrochloric or sulphuric acid solution (Platinum is precipitated by hydrogen sulphide from acid solution over a wide range of acidity).

(*8*) Attack of boiling sulphuric acid on platinum is almost entirely prevented by the presence of an excess of sulphur dioxide (generated by the interaction of the hot acid and sulphur or carbon, or by previously adding sulphurous acid containing 6% w/v SO_2).

(*9*) The solvent action of boiling perchloric acid on platinum is negligible.

(*10*) When heating some platinum ware above 1,100°C a slow loss in weight may occur; this loss increasing with the temperature is usually described to volatilization of iridium metal which is alloyed with platinum for hardening effect. When very long ignitions at high temperatures in platinum vessels must be made of substances whose exact weight is to be determined, a correction must be applied and the loss in weight of the vessels should be ascertained after removing the ignited residue without injuring the vessels, or else (when the loss of weight of the substance is of no consequence) the weight of the vessels should be taken after and not before ignition of the substance.

(*11*) Platinum vessels should be clean and burnished. Traces of insoluble residues may ordinary be completely removed by gently fusing with some crystals of bisulphate for a few minutes and then digesting the vessels in hot HCl (1 + 1); some residues require other fluxes and acids. Dull vessels should be burnished with wet sand (–100 mesh to the linear inch) applied with the fingers.

(*12*) Preferable are platinum crucibles reinforced by strong platinum wire on the rim and provided with well-fitting platinum covers. Crucibles of 10-ml capacity may be satisfactory for small quantities; those of 20–25 ml capacity are used for most purposes as for ignition and weighing of numerous materials, and for fusions which must be made in platinum. Flat-bottom reinforced platinum dishes of 100–500 ml capacity having a small lip to aid in pouring and provided with well-fitting covers may be used in accurate determinations of silica, or when alkaline or neutral solutions have to be evaporated or must stand over any length of time. (Round-bottom dishes should not be used, because there is a danger of local overheating and baking in evaporating to dense fumes of acid, or to dryness.)

Substitutes

In many operations most of the platinum ware listed above can be replaced by ware of other materials to good advantage, viz. by:

Quartz

Great resistance to attack by acids (excepting HF), slightly hygroscopic, small coefficient of expansion, at 1,200°C less volatile than platinum, softening point 1,470°C, boiling point 2,230°C, material for: crucibles for the ignition and weighing of substances and for pyrosulphate fusions in which silicon is not to be determined, dishes for dissolution of acidic smelts and for evaporating of acidified solutions, Gooch crucibles with fixed bottoms, triangles, gastight combustion tubes, etc.

Porcelain

Also a substitute for glass (see p. 5), material glazed or unglazed, for: gastight combustion tubes, crucibles for the weighing and ignition of substances at low temperatures and for special fusions with sodium peroxide.

Nickel, iron and zirconium

For crucibles of 30 ml capacity with covers needed in fusions with sodium peroxide made of as pure metal as can be obtained, the metal depending on the end in view.

Gold and silver

For crucibles of 30 ml capacity with covers to be used in certain fusions with potassium or sodium hydroxide; their low melting points (1,050°C for gold, and 950°C for silver) must be taken into account.

REAGENTS

Unless otherwise specificated, all reagents mentioned in this book shall conform in purity with the recommended specifications for analytical reagent chemicals (see e.g., "Analar Standards for Laboratory Chemicals" The British Drug Houses, Ltd., and Hopkin and Williams, Ltd.).

When the reagents are specified by name or chemical formula only, it shall be understood that concentrated reagents of the specific gravities or concentrations shown in Table I are intended.

In expressions (1 + 2), (5 + 3), etc., used in connection with the

TABLE I

Strength of the common acids and ammonium hydroxide

Reagent	*Formula*	*Approximate*	
		specific gravity	*percentage of reagent by weight*
Acetic acid	CH_3COOH	1.05	99.5
Hydrobromic acid	HBr	1.38	40
Hydrochloric acid	HCl	1.19	37
Hydrofluoric acid	HF	1.15	46
Nitric acid	HNO_3	1.42	70
Perchloric acid	$HClO_4$	1.67	70
Phosphoric acid	H_3PO_4	1.69	85
Sulphuric acid	H_2SO_4	1.84	96
Sulphurous acid	H_2SO_3	1.03	6 (SO_2)
Ammonium hydroxide	NH_4OH	0.90	27 (NH_3)

name of the reagent, first numeral indicates volume of reagent used, and second numeral indicates volume of water, as in the following example: HCl (5 + 95) means 5 volumes of concentrated HCl (s.g. 1.19) diluted with 95 volumes of water. The term "water" means distilled water.

In making up solutions of definite percentage it shall be understood that x g of anhydrous substance is dissolved in water and diluted to 100 ml.

In the case of certain reagents, the concentration is specified as a percentage by weight, for example: HCl (10%) means a solution containing 10 g of HCl per 100 g of solution (cf. Table I).

Concentrations of standard solutions are expressed as normalities or as equivalents in mg per ml of the element to be determined, for example: HCl (0.1 N), silicon solution (1 ml ≡ 0.050 mg Si), $KMnO_4$ solution (1 ml ≡ 0.5595 mg Fe).

FILTERS

Strong filter paper of at least three textures, hydrochloric-hydrofluoric-acid-washed, of uniform texture and known low ash content, should be at hand, as, e.g., Whatman's "ashless" filter paper No.40 for most crystalline precipitates, No.41 for gelatinous precipitates, No.42 for finest precipitates and precise work and for use with Buchner funnel and vacuum pump. Sizes may vary from 4 to 25 cm diameter; 9 cm and 11 cm (cubic capacities of conical filter with angle of 60°, respectively: 20 ml and 37 ml) being most useful in macro analyses.

Filter paper may contain soluble organic matter of slightly reducing nature which, if required, shall be removed before filtration of oxidized solutions as e.g., in the determination of small amounts of chromium by colorimetric comparison with standards after oxidation in alkaline solution to the sexivalent stage, by which the filter paper should first be washed thoroughly with a solution of sodium hydroxide (50 g NaOH per liter) in order to remove soluble organic matter, thus preventing reduction of the chromium to a lower stage giving a false colour in the colorimetric determination. Another example of reduction is shown in the volumetric determination of vanadium, based on titration with standard permanganate after reduction by sulphur dioxide or ferrous sulphate, by which too high results may be obtained caused by the extraction of oxidizable matter from the filter paper. In such cases reduction can be entirely avoided by filtration through asbestos or other inorganic material (see below).

Macerated filter paper of known low ash content, like Whatman's "ashless" filter paper tablets, is recommended in the final filtration of gelatinous precipitates; here the paper pulp should be added after precipitation has taken place, never before!

Filter funnels should have a regular slope of 60°, a length of stem not more than 15 cm and an internal diameter of 3–5 mm; for more rapid filtration of colloidal precipitates like that of the ammonium hydroxide group, fast running filter papers and Jena funnels with exceptionally long stem of very small internal diameter are available.

Fritted-glass filtering crucibles and -funnels of chemical-resistant glass, or porous-bottom porcelain filtering crucibles, and suction filtration, are recommended when the precipitate is only to be dried at low temperatures or filtration has to be done through inorganic material; available in different sizes and grades (5–240 microns) of porosity, below 150°C crucibles and funnels made of glass are to be used, above 150°C crucibles of porcelain.

Gooch crucibles of different capacities (10 ml, 20 ml, 25 ml, and 35 ml) with fixed or removable bottoms made of platinum, porcelain, or fused silica, in combination with a suction arrangement may also be used; the filter pad may consist of purified asbestos fiber resistant to chemical attack, of a mat of silica cotton, or other suitable inorganic material.

Asbestos of chemical stability and a satisfactory filtering speed to be used for Gooch crucibles should consist of three different sizes of a good grade of white asbestos being purified as follows: digest three sizes (coarse, medium, and fine fiber) of Gooch grade anhydrous, white, silky, shredded, asbestos of amphibole variety, in three separate covered beakers of Pyrex glass in hydrochloric acid on a steam bath for several hrs, filter the three sizes separately on a Büchner funnel, wash thoroughly with water and repeat the acid digestion and washing until all soluble iron and chlorine are removed. (Sensitive and reliable tests on the complete removal of soluble iron and chlorine: when 2,4-dinitroresorcinol, $C_6H_2(OH)_2(NO_2)_2$, is added to some ml of the acid filtrate or washings, a yellow colour indicates incomplete removal of either Fe''' or Fe''; when a few drops of 0.1 *N* solution of $AgNO_3$ are added to some ml of the filtrate or washings, a white curdy precipitate of AgCl, insoluble in nitric acid, but entirely soluble in ammonia, indicates incomplete removal of Cl^-.)

If the asbestos is to be used for a particular operation, further purification by treatment with other reagents and solutions, or preliminary drying or ignition at the same temperature as that in the analysis, may be necessary. After preparation the three sizes of asbestos fiber should be stored separately in wide-mouth bottles.

CHAPTER 3

Laboratory Instruments

pH METERS

An adjustment of the alkalinity or acidity of a solution is necessary in some methods and this may be achieved using either a pH meter, indicator solutions or indicator papers of various ranges; the latter should be stored in a clean atmosphere and only the minimum requirement left in the laboratory. Indicator solutions should be kept in the dark in well-stoppered bottles of resistant glass.

Care of the electrodes is the most important point of maintenance of a pH meter, the electronic circuits usually being enclosed do not require attention. Most models use a mains supply but a few portable models require a dry battery as a source of voltage for balancing the potentiometric circuits.

When the instrument is not in use the electrodes should be stood in a slightly acid solution, e.g., HCl (1 + 99). The electrodes should never be left dry. The reference electrode must be filled to the required mark and the instrument set up on two buffer solutions before use, one at a low pH about 4 and another at about 9. If an automatic temperature control is fitted, the switch should be set to this, otherwise the manual control must be set to the temperature of the solution.

At no time should the glass electrodes be placed on the bench top as small scratches will damage the thin glass envelope.

SPECTROPHOTOMETERS

Photometric methods are generally based on the measurement of the transmittancy or absorbancy of a solution of a coloured salt, compound, or reaction product of the substance to be determined. When light of a definite wave-length or restricted wave-band (which may be in the ultraviolet and near-infrared region of the spectrum) is employed, the technique is known as spectrophotometry. Although sought elements, compounds, or ions may themselves have colour, it is more usual to introduce colour into the system by adding a reagent. Inorganic reagents function by oxidizing, reducing or complexing; organic reagents bring about adsorption, oxidation, reduction, fluorescense or chelation. Ideally the colour of the system should be rapidly produced at room temperature by means of a colourless specific reagent, and it should be intense and stable, not too responsive to pH changes, and should conform with the Lambert-Beer law. Since the measurement is made by comparing the ratio of light incident to the light after passing through a fixed path length of solution, the term "absorbance" is used.

In the early days of the development of colorimetric techniques, a visual comparison was made. Visual colorimetric matchings may be effected, e.g., by: adjustment of the depth of test solution or standard (if the solution of the coloured compounds obeys Beer's law), treatment of comparison blank (subsequently brought to the same volume as the test) with a standard solution, comparison with a series of liquid standards of equal volume or with standard glasses as in comparators and tintometers. Visual colorimetric methods are still widely used when simple, rapid procedures of moderate accuracy answer the question at hand, but they generally cannot be used when other coloured compounds are present. The most common visual method is that in which the sample and reference standard solutions are matched in Nessler tubes. The precision of the measurement can be greatly improved if a colorimeter (such as Duboscq or Pulfrich) together with suitable light filters is used; with an instrument of the Pulfrich type it is possible to obtain a calibration curve using solutions of known concentration and thus eliminating the necessity, during an analysis, of direct comparison against reference standards. Photometers employing filters are known as spectrophotometers and usually isolate relatively broad bands of light.

Spectrophotometers are designed to compare the absorbance of solutions by measurement of the light after passing through the solution either by the intensity falling upon a photocell or by electronic multiplication of the signal emitted by a photocell. The intensity of the light source is required to be constant and is controlled by a variable slit. In practice the intensity of the light is adjusted by placing water in the absorption cell and closing the slit until zero absorbance is registered by the meter. Between measurements this zero is checked to guard against a change in emission of the light source. If this does vary it usually indicates that the current supplied to the lamp has altered. A constant emission is usually reached after the instrument has been in operation for several minutes.

The simplest instrument using a galvanometer to measure the current produced by the photocell, is shown in Fig.1. The construc-

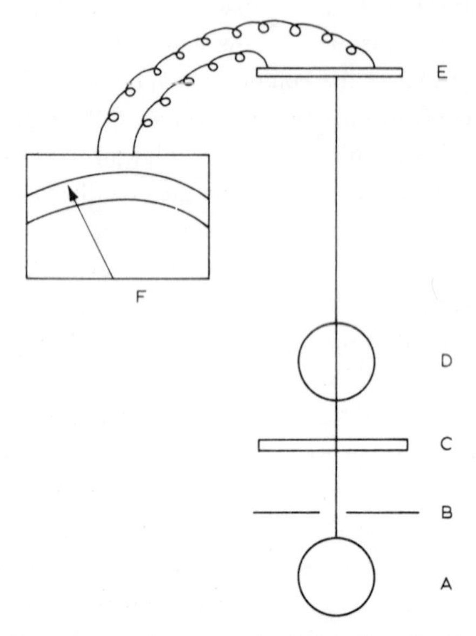

Fig.1. Simple filter spectrophotomer. A = lamp; B = slit; C = filter; D = cell containing solution; E = light-sensitive cell; F = galvanometer.

tion used in a number of the currently available models has been described in some detail by Vogel (1958, p.611).

Coloured solutions show a maximum absorbance in a particular portion of the spectrum and in a simple instrument this portion is selected by means of a coloured filter (Table II). In the more complex instruments a quartz prism is used to select that portion of the spectrum in which the maximum absorbance occurs (Fig.2). A smaller portion of the spectrum may be selected using a prism, and this allows a more selective measurement to be made at the peak of the absorbance.

A number of different spectrometers are available, the choice being controlled by the degree of sensitivity required and the finance available. The cheaper models using filters and a direct reading of the galvanometer are the simplest to use whilst the more expensive models, having enhanced sensitivity, are more suitable for the determination of trace quantities of constituents. The wavelength required for each of the spectrophotometric methods is given under the appropriate procedure in the text. In Table II, suitable filters for use with the EEL spectrometer are given

The glass cells used to contain the solutions during measurement require careful handling to prevent the optical surfaces from becoming scratched and damaged. Since some complexes precipitate upon

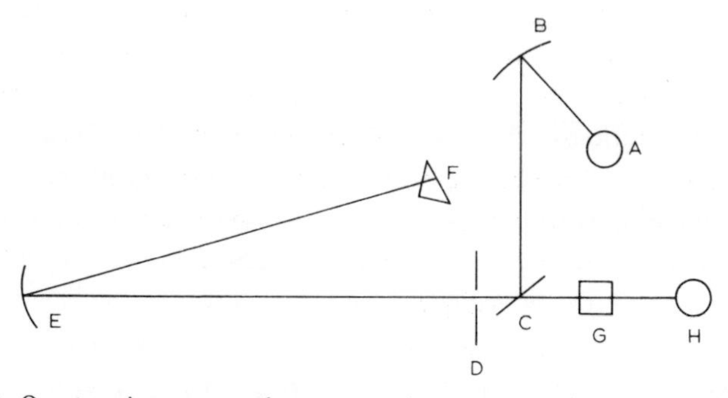

Fig.2. Quartz prism spectrophotometer. A = tungsten lamp; B = condensing mirror; C = diagonal mirror; D = slit; E = collimating mirror; F = aluminized quartz prism; G = cell containing solution; H = photocell.

TABLE II
Suitable filters for the use with the EEL spectrophotometer

Constituent	*Wavelength*	*EEL filter*
Silica	640	607
Aluminium	410	600
Total iron	522	604
Ferrous iron	522	604
Manganese	525	604
Titanium	400	600
Titanium	430	601
Chromium	370	600
Chromium	540	605
Phosphorus	650	608
Phosphorus	430	601

standing, after the measurement of the absorbance is complete the cells should be washed several times with distilled water and stored in a polyethylene beaker containing HCl (2 + 98). If however precipitation does occur inside the cell, this should be removed with a wet swab of cotton wool wound around a wooden splint. Continual use of strong acids, e.g., chromic acid, is not recommended as slight etching of the optical surface may occur.

In some instruments a carriage is fitted allowing two cells to be placed alternately in the path of the photocell. One cell, filled with water, is reserved for reference, the other is used for the sample and standard solutions. Where possible the purchase of matched cells is recommended, since before commencing a measurement the cells to be used are filled with water and a comparison made of their absorbance. The difference in absorbance between the reference cell and the cell used for the sample solution is known as the cell blank. Using the cell blank a correction is made to the absorbance of the sample solution. When matched cells are constantly in use, a larger than usual cell blank indicates that one of the cells requires cleaning. When a wider than normal slit width is required to zero the instrument with water in the reference cell, this is another indication of dirty cells.

If wet batteries are used to supply the current for the instrument, these should be kept filled to the required mark with electrolyte and fully charged. In some instruments internal dry cells are used in the electronic circuits. Although this type of battery has a good discharge life, they should be checked for voltage if sensitivity is less than normal or at six monthly intervals depending upon the use made of the instrument.

FLAME PHOTOMETERS

The main application of these instruments in silicate analysis is in determination of the alkalies sodium and potassium, although its use may be extended to other elements. The instrument may be either an attachment, e.g., Beckman flame attachment to the D.U. model spectrophotometer or an independent instrument.

The principle of flame photometry is that sample solutions to be analysed are atomized into a flame and light, characteristic of the elements to be determined, is isolated; the light intensity (the radiant energy at characteristic wavelengths) is measured against that of atomized portions of standard solutions. As with spectrophotometers the measurements may be made either by measuring directly on a galvanometer the current produced by a photocell or by electronic multiplication of the signal received by a photocell.

An important part of the instrument is the injection system through which the solution is drawn into the flame. At some point in all of the various designs, the solution passes through a very fine orifice. It is essential for accurate measurement that this orifice is not allowed to become blocked. Fine particles precipitated from the sample solution, e.g., calcium sulphate, or small particles of filter paper if the solution has been filtered, will cause a blockage. Filtration through hardened filter papers (e.g., Whatman 500 series) will reduce the free particles from the paper surface. If however blockage does occur only the very fine wire supplied by the manufacturer should be used to clear the orifice. Where a jet is detachable, e.g., Beckman attachment, it may be periodically cleaned by sucking through the exit end, with a vacuum pump, a cleansing solution such as dilute acid solution containing a general complexing agent, e.g., EDTA. Distilled water is sucked through before the jet is returned to operation.

The combustible gases should be free of contamination by particles of dust, oil or water to ensure an evenly burning flame. Where an air compressor is used, provision should be made for the inclusion of a filter unit to remove any contaminants.

As with spectrophotometers a constant source of current must be made available for the electronic circuits.

CHAPTER 4

Major Factors in Spectrophotometric Methods

CHOICE OF METHOD

Numerous papers on analytical chemistry are currently being published and are readily available to students in institute libraries. A number of these papers give the basic conditions necessary for a particular reagent to react with the desired constituent leaving the detailed work to be completed by interested analysts. Other publications give an application of the reagent in a different field of interest which may entail considerable work in its adaption to silicate analysis.

New reagents sometimes become popular and their application to the various sections of analysis is pursued with intensity: dithizone, diethyl dithiocarbamate and EDTA have all passed through this stage. In time some of these methods become incorporated into the general practice of analytical chemistry but many are rarely used and then only for special applications. Because of this it is necessary for the student investigating current literature to be most wary of new and little tried methods as a great deal of time may be wasted investigating and amending the method to the requirements of silicate analysis.

Since the factors which affect spectrophotometric methods are less appreciated than those in gravimetric analysis these factors are given below with comments on their application to silicate analysis. In most cases more than one single factor is critical in a method, for example in the spectrophotometric determination of silicon; temperature, pH of solution, strength of molybdate solution and time of reaction all influence the final value of the absorbance.

In general, spectrophotometric methods are based on two types of reactions: viz., coloured complex formation and coloured lake formation. The latter is generally more difficult to standardize since slight changes in the technique used will have a large effect on the colour developed. Because of this it is desirable to perform a rather complete spectrophotometric investigation of the colour reaction before attempting to employ it in quantitative analysis. The investigation should include a study of the following: (*a*) The specificity of the colourforming reagent; (*b*) the validity of Beer's law; (*c*) the effect of salts, pH, temperature, concentration of the colourforming reagent, and the order of adding the reagents; (*d*) the stability of the colour; (*e*) the absorption curve of the reagent and the coloured substances; (*f*) the optimum concentration range for quantitative analysis.

The major factors which influence spectrophotometric methods are as follows: (*1*) pH of solution; (*2*) oxidation and reduction of solutions; (*3*) presence of foreign ions; (*4*) period of reaction; (*5*) stability of complexes; (*6*) reagents; (*7*) light sensitivity of the complex; and (*8*) temperature.

pH of solution

The pH of the sample solution during the formation of the complex may control the absorbance of the solution and therefore requires careful attention. An example of this is the formation of the α and β forms of the silicomolybdate complex, in which the change from one species to the other is not immediate and the predominant form is controlled by the initial pH of the solution.

A different type of behaviour is shown by the titanium tiron complex. This complex forms one species at a pH of 2–4 which has constant absorbance within this range of pH and another species at pH $>$ 4, which also has a constant absorbance but at a higher level. The change from one species to other in this case is immediate upon changing the value of the pH (Nichols, 1960).

In some cases the absorbance of a complex changes continuously over a certain range of pH after which the absorbance remains constant. The dipyridyl ferrous iron complex which has a zero absorbance at pH 2 and increases to a maximum at pH 5.5 is an example of this.

From the above illustrations it will be seen that the pH value of both the original and final solution may be critical in the formation of coloured complexes. Where a strict control of the pH of the original solution is not possible, e.g., after a HF-H_2SO_4 treatment of the sample, an attempt must be made during fuming to keep the residual sulphuric acid constant.

Adjustment of pH values of solutions may be monitored by using a pH meter, indicator solutions or indicator papers.

Oxidation and reduction of solutions

Oxidation or reduction of solutions is comparatively simple but the change of state is seldom instantaneous and therefore sufficient time must be allowed for the reaction to be completed. Where an oxidation is required, it does not necessarily follow that one oxidant may be replaced by another chosen at random. For example, in the oxidation of chromium, bromine water will not raise the chromium to the state of chromate necessary to facilitate its separation from iron by the neutralisation of the solution using sodium carbonate.

When it is required to oxidize manganese to the permanganate state for measurement of the absorbance, one of two oxidants is usually employed. Potassium periodate has a slight advantage in that it may be used in the presence of small traces of chloride. However, the use of a mixed solution of mercuric sulphate, phosphoric and sulphuric acid (Riley, 1958a) will allow potassium persulphate and silver nitrate (catalyst) to be used in the presence of traces of chlorides. Oxidation using the latter reagents is slightly quicker and appears to be more complete.

It is possible that dust falling upon the surface of a dilute oxidized solution will cause reduction.

Difficulty is experienced in retaining large quantities of iron in the ferrous state. This, however, is often required in order to avoid interference with a complex during measurement. An illustration of this is shown in the reduction of the iron tiron complex during measurement of the titanium tiron complex. Sodium dithionite is used for the reduction for two reasons. First, only a small quantity of the reagent is required to reduce quite large amounts of iron and secondly the liberated SO_2 forms a gaseous blanket over the surface of the solution thus reducing air oxidation. If the reducing agent is added while the pH of the solution is 2, no coloured complex is formed.

Some compounds are more easily reduced than others, for example EDTA is able to reduce permanganate solution within a few minutes but a period in excess of 30 min is required to reduce chromate solutions. This effect has been used to eliminate interference of one element with the other.

From the above it is clear that due attention must be paid to the use of the correct oxidizing or reducing agent and to the pH value of the solution.

Presence of interfering elements

When the photometric measurement in the sample solution is made without previous separations, it usually happens that one or more of the elements or reagents present may interfere with the colour reaction; the interference may be due to the presence of a coloured substance, to an enhancing or suppressive effect on the colour of the substance being measured, or to the formation or destruction of a complex with the colour-forming reagent thus preventing full colour development. Foreign elements may interfere by means of one or more of the following mechanisms: (*a*) consumption of the reagent; (*b*) formation of a similar coloured complex; (*c*) formation of a different coloured complex; and (*d*) restriction of a reaction. In the following paragraphs each of these will be examined in turn, giving examples having a direct bearing on silicate methods.

(*a*) There is danger in the use of only a moderate excess of the reagent where other constituents in the solution also react with the reagent, thereby complexing it preferentially and restricting its reaction with the desired constituent. Where tiron is used to determine titanium the reagent is also held by aluminium and calcium which form colourless complexes. It is necessary therefore, to add sufficient tiron to ensure that an excess is left to combine with the titanium. For this reason 100 mg of tiron is added although less than 1% of this weight is actually required by the titanium. The investigation of interference caused by other elements in the sample solution may prove essential to the effective utilization of a method as is clearly illustrated in the above instance.

(*b*) A common form of interference is the formation of a similar coloured complex since most reagents are not specific for the constituent that is being determined. It is to counter this interference that complexing agents are added to restrict the combination of foreign elements with the reagent. Iron is one of the most common constituents of the sample solution and this may be complexed by a number of reagents, e.g., citric acid, EDTA and phosphoric acid; the latter is used to complex iron when titanium is determined as the peroxide complex. Where iron in the ferrous state does not interfere, the addition of a reducing agent is the most common method of reducing interference of ferric iron.

(*c*) Where a coloured complex is formed by the interfering elements, with a maximum absorbance at a different wavelength, it is sometimes possible to measure the absorbance at this wavelength and to estimate the extent of the interference to be deducted. Methods have been published using this technique (Burke and Yoe, 1962) where more than one constituent may be measured in a single solution.

When manganese is determined as permanganate, chromium if present in large quantities, e.g., $>3\%$ by weight, will interfere. The yellow chromate colour when superimposed upon the pink of the permanganate will give an orange colour. The only recourse in this case is to measure the combined absorbance and then to destroy the less stable, i.e., $KMnO_4$ with EDTA and measure the residual chromate absorbance which is then deducted.

(*d*) During the determination of one constituent, it is possible that another will form either a soluble or insoluble complex which will interfere with the required reaction. If this second constituent forms a stronger complex than the reagent a low result will be obtained. By careful choice of reagent this interference may sometimes be eliminated. Interference of phosphate in the determination of iron using potassium thiocyanate is an example of this; when dipyridyl is used this interference is eliminated.

Period of reaction

Some spectrophotometric methods use reactions in which the coloured complex develops rapidly, e.g., dipyridyl with ferrous iron and hydrogen peroxide with titanium. In these reactions, the maximum absorbance is obtained almost immediately upon mixing the reagents with the sample solution. Other spectrophotometric methods use reactions in which the maximum absorbance is not reached until the solution has been standing for some time. In these latter cases the absorbance is measured after a given interval of time has elapsed from the addition of the reagent. In general absorbances of standards and samples should be measured during the time interval of maximum stability of the colour, provided that this occurs reasonably soon after development of the colour. In those cases where the colour changes continuously and never reaches a stable intensity (as in the spectrophotometric determination of silicon, p. 81), the reaction is stopped after a given interval by the addition of another reagent and a uniform time for colour development should be used for both calibration solutions and samples, the interval measured with the aid of a stop clock.

In methods where a coloured lake is produced, the rate of addition of the reagent generating the lake is usually critical. To obtain reproducible results the addition of the reagent must be controlled at a reproducible rate, i.e., a specified number of drops/min.

Stability of complexes

Unless a specified time interval is given in the method, the absorbance of a solution containing a complex should be measured immediately after development. The determination of chromium using diphenylcarbazide is a case in which this is of extreme importance; the colour of the chromium complex develops almost immediately but starts to fade within a short period of time. After the development of the complex, a pure chromium solution is stable for a considerable time but oxidizing agents cause rapid fading of the chromium complex.

Reagents

The exact quantity of the reagent which is added to develop the coloured complex, is not critical in a number of methods provided sufficient of it is present. However, it is advisable to adopt the practice of accurately measuring the quantity of each reagent. Care should also be taken, particularly when adopting a method from a different branch of chemistry, to check that sufficient of the reagent is added to complete the reaction. For example, methods used in the testing of water may be concerned with microgram quantities whereas in silicate analysis milligrams of the constituent may be present in the solution. This would entail the use of a larger amount of the reagent.

A number of organic reagents tends to decompose on storage and larger quantities of the material are required to be added therefore to obtain the required strength of solution for the reaction. The date of receipt should be clearly marked on each container when organic reagents are bought into a department.

Photo-sensitive complexes

Solutions of coloured complexes which are decomposed by the action of sunlight (like the dimethylglyoxime complex or the cobalt complex formed by the nitroso-R-salt method) should be kept in the dark during the time required for full development of the colour before measuring.

Temperature

Some reactions used in spectrophotometric methods have a temperature coefficient which is sufficiently high to introduce an error in the absorbance measurement. The yellow unreduced silicomolybdate complex is an illustration of this behaviour. In such methods the flasks containing the reacting solutions should be left standing in a bath of water at a fixed temperature, preferably slightly in excess of normal room temperature to allow for possible increases of temperature in summer time. This procedure is essential where a reaction is sensitive to temperature and it is desired to use a standard curve. In certain photometric methods the temperature coefficient of the colour reaction is large enough to require winter and summer standard curves or charts.

CHAPTER 5

Common Operations in Silicate Analysis

DECOMPOSITION BY FLUXES

The following commoner fluxes are used to obtain solution of silicate material, the choice of fluxes and their containers depending upon the end of view.

(*a*) Sodium carbonate (as most frequently used flux); an equimolecular mixture of sodium- and potassium carbonate (because of its lower m.p. when, e.g., chlorine or fluorine is to be determined); a mixture of sodium carbonate and potassium nitrate (where an oxidizing fusion is required, KNO_3 to an amount little greater than needed to oxidize fully the oxidizable components of the sample as, e.g., small quantities of sulphides). Fusions with the fluxes mentioned under (*a*) must preferably be made in platinum. Consult: platinum and substitutes (pp. 6–9).

(*b*) Potassium- or sodium bisulphate (for oxidized substances, e.g., for dissolving ignited precipitate mixtures in the course of analysis with a view to the determination of their components exclusive phosphorus). Fusions must preferably be made in platinum under the same precautions mentioned under (*a*) by gently heating at the start and at a temperature not higher than is required to maintain liquidity because platinum is somewhat attacked by pyrosulphate fusion (see p. 7).

(*c*) Sodium peroxide (more used to bring certain elements to the required state of oxidation and to separate them from others than in making complete analyses). Generally, fusions must be made in iron, nickel or zirconium (see p. 9).

Sodium carbonate fusion

(*1*) Decomposition of the sample. Ignite a clean platinum crucible of 25 ml capacity with well-fitting platinum cover grasped with platinum-tipped crucible tongs above the blue or green inner zone of the non-luminous flame from a Bunsen- or Meker burner for some minutes and place crucible and cover in a desiccator above colourless sulphuric acid. As soon as cool, place crucible by means of the crucible tongs in the center of the left pan of a sensitive analytical balance and weigh the empty crucible against weights of good class transferred from the box to the right balance pan with ivory-tipped forceps.

For the main portion (see Scheme I, p. 85) transfer 1 g of the finely ground air-dry sample to the crucible and reweigh accurately. Add about 4–6 g of anhydrous sodium carbonate, mix carefully for at least 5 min using a platinum rod, clean the rod of adhering particles by scraping with another rod, and cover the mixture with a layer of about 2 g of anhydrous sodium carbonate. Cover the crucible, heat on a platinum (or quartz) triangle over a low flame from a Bunsen burner to expel any moisture in the contents, direct the flame at an angle against side and bottom of the crucible, and increase the flame gradually to full heat (approximately 1,000°C) avoiding violent action and loss by boiling over by lifting cautiously the crucible cover at intervals. When the contents are in quiet fusion, gently rotate the crucible by holding it with a pair of platinum-tipped tongs to stir up any unattacked particles and heat till an almost viscous liquid is obtained, and,, when heated over a Meker burner (approximately 1,200°C) no further effervescence occurs. Undecomposed particles adhering to the cover should be fused over the free flame till a clear fusion drop is obtained preventing the molten drop from running from the cover.

(*2*) Removal of fusion cake from the crucible. The fusion cake may be removed by one of the following methods.

(α) Before the melt solidifies, remove the cover, place the bent end of a short platinum rod in the crucible and place the crucible on a thick marble block to cool. When cool, remove the fusion cake by lifting the platinum rod from the crucible. When the cake does not detach itself, add a few ml of hot water and allow to stand on a hot steam bath for 10 min before lifting the cake from the crucible.

(β) While it is still molten incline the crucible so that the melt runs slowly towards the edge and hold the crucible in this inclined position until the melt has solidified. When quite solid, cool the side of the crucible that is opposite to the melt under a thin stream of cold water; this will cause the platinum to contract away from the melt. When this side of the crucible is quite cold direct the stream onto the opposite side. The fusion cake should now separate from the surface of the crucible. A slight tap is usually sufficient for the major part of the fusion cake to come away. Small remaining portions can be removed by the addition of a few ml of water to which has been added one or two drops of alcohol and hydrochloric acid.

The ease with which a fusion detaches itself from the platinum largely depends on the smoothness of the inside surface of the crucible. After a number of fusions has been made, this may require repolishing by rubbing a very smooth pencil of wood over the surface until the polish is again present.

Potassium bisulphate fusion

When the determination of phosphorus is of no consequence this fusion is mainly used for solution of the ignited ammonium hydroxide group and for minerals remaining after treatment of the sample with hydrofluoric and sulphuric acid. The main difference between a sodium carbonate and a potassium bisulphate fusion is that the reaction in the latter occurs at a much lower temperature, namely just below red heat. Besides this in its initial stages the latter must be watched closely for much water is given off accompanied by frothing and substance is spattered onto the crucible lid. The major part of the fusion occurs when the bisulphate has been converted to pyrosulphate, however some of the more refractory oxides may still remain. If the fusion is still not completed by the time the pyrosulphate has decomposed leaving a sluggish potassium sulphate melt, the melt should be allowed to solidify, one drop or more of sulphuric acid should be added, the melt heated cautiously to restore the flux to its effective state after which the crucible is reheated at a low temperature until all portions of the melt are fluid again. Fusion is then continued to completion and the crucible allowed to cool to room temperature. (This cooling before the addition of an aqueous solution is most necessary for reasons of safety.) The fusion cake is dissolved in H_2SO_4 (1 + 9) by carefully heating the crucible over a low flame until a clear solution is obtained. This solution may contain some mg of dissolved platinum because platinum is somewhat attacked by pyrosulphate fusions (see p. 7).

Sodium peroxide fusion

Sodium peroxide is very reactive. It causes a much higher rate of corrosion of crucibles normally used for the fusion of silicate material; the speed of attack rapidly increases with temperature. The loss in weight of a platinum crucible at 550°C using 5 g of sodium peroxide could be as high as 20 mg (Belcher, 1963). For this reason crucibles made from iron, nickel or zirconium (depending on the end in view) should be used, the latter being the more expensive (see p. 8). Iron and nickel crucibles have a short life because they are strongly attacked by sodium peroxide. If it is essential that iron and nickel must be absent, a zirconium crucible may be used with an expected contamination of 0.1–1.0 mg Zr. Since chromium is only expected to be present in zirconium to the extent of a fraction of one per cent, the contamination from the crucible during the fusion of chromite will be very low. Crucibles made from pure ingot iron contain only traces of silicon and the amount of SiO_2 present in the Na_2O_2 used is usually negligible but blanks should be run on new batches of crucibles and new lots of Na_2O_2. It is advisable to use the best grade of sodium peroxide obtainable, high purity grades, e.g., B.D.H. "micro-analytical use" are available.

Procedure

Transfer about 8 g of dry yellow sodium peroxide to a 30- or 50-ml crucible made from the proper material, add 0.5–1.0 g of air-dry sample (–200 mesh), mix well using a platinum rod, carefully clean the rod of adhering particles by scraping with another rod, and cover the mixture with a layer of about 2 g of sodium peroxide. Heat the crucible and contents on a hot plate for 10–15 min to expel any water in the sodium peroxide that would cause spattering in the subsequent fusion. Carefully fuse over a low flame by holding the crucible with a pair of iron tongs and slowly revolving it around the outer edge of the flame until the contents have melted down quietly, taking care that the reaction will not proceed so violently as to cause spattering. When the flux is molten, rotate the crucible carefully to stir up any unattacked particles on the bottom or sides, the crucible and contents being rotated and maintained at a low red heat (600°C) for about 5 min and finally to bright redness (700°C) for 1 min. (It has been shown (Rafter, 1950) that minerals such as quartz, talc, rhodonite, beryl, chromite and ilmenite are decomposed $<300°C$, rutile and even zircon, a most resistant mineral, $<400°C$.) Cover the crucible, allow to cool almost to room temperature, and tap on an iron plate to loosen the fused mass in a cake. Transfer the cold cake to a dry platinum dish, cover with a platinum lid, and cautiously add 50 ml of water. When the reaction ceases, wash any small amount of material adhering to the crucible into the dish with a little water. Cool the solution and add 25 ml of H_2SO_3 (6% wt./v SO_2) to prevent attack on the dish by chlorine upon acidifying with hydrochloric acid. Cool, and carefully add hydrochloric acid until in moderate excess. If a platinum dish is not available for solution of the fused cake, it may be disintegrated with water in a pure nickel dish and the contents then transferred to a porcelain dish (of good glaze) containing sufficient hydrochloric acid to provide an excess of acid. It is not desirable to dissolve the fusion directly in porcelain because of the action of the alkaline solution. If porcelain or pyrex must be used, it is necessary to carry along duplicate blanks. However this method fails if substances that are present in the blank do not undergo the same reactions as those in the test (see Glassware and Porcelain, p. 5).

Fusion of sulphide-containing minerals

If the sulphide content of the sample is low, 1 part of the sample should be mixed thoroughly with 10 parts of a mixture, consisting of 10 parts of anhydrous sodium carbonate and 1 part of potassium nitrate, and covered with a layer of the flux. Potassium nitrate will oxidize sulphide to sulphate but its amount should be kept to the minimum to save the platinum crucible (see *4* p. 7). The initial period of heating should be carefully controlled as the liberation of gases will give rise to frothing. During the fusion the flame should be directed at an angle against side and bottom of the crucible to prevent a reducing atmosphere within the crucible. Without special precautions dissolution of the melt in hydrochloric acid should never be done in contact with platinum (see *5*, p. 7).

Samples containing a higher oxidizable content should be treated as follows. Transfer approximately 1 g of the sample (–100 mesh) to a 250-ml pyrex beaker, add carefully 20 ml of bromine water and 20 ml of hydrochloric acid, and cover the beaker with a well-fitting pyrex watch glass. Let stand at room temperature with occasional shaking for half an hour, next transfer beaker and contents to a steam bath, heat until all action has ceased, place the watch glass on bent glass rods, and evaporate to dryness. Let cool, moisten the dried mass with 5 ml of hydrochloric acid, after 5 min add 100 ml of hot water, rinse the watch glass, rods, and inside of the beaker with hot water, replace supports and watch glass, boil gently for 5 min, filter through an ashless Whatman filter paper No.42 into a porcelain dish (of good glaze), wash beaker, paper and residue, five times with hot water and finally transfer the residue completely to the paper. Ignite paper and residue in a platinum crucible, fuse with anhydrous sodium carbonate, dissolve the melt in an excess of hydrochloric acid, add the solution, if clear, directly to the reserved filtrate and evaporate both for the silicon determination. If particles of a finely ground sample pass through the filter paper into the filtrate, the residue after a HF-H_2SO_4 treatment of the silica will be slightly higher than usual.

DECOMPOSITION BY ACIDS

When silicon is not to be determined, the main purpose in preparing a solution of the sample by treatment with hydrofluoric acid is to obtain a solution free of the major element silicon. This solution may also be used for the determination of some constituents by spectrometric and flame photometric methods (see Chapter 17).

In addition to hydrofluoric acid, the following acids, single or in combination, are added in minor proportions: sulphuric, perchloric, nitric or hydrochloric acid. The reason for this is that aluminium which is commonly present in silicate samples, forms a fluoride which when heated attacks the platinum dish. Any of the four acids given above decomposes aluminium fluoride and prevents this attack. Moreover, it is impossible to precipitate aluminium completely by ammonium hydroxide unless the last traces of fluorine are expelled which can be effected by evaporating with either perchloric acid to dryness or by evaporating twice with dilute sulphuric acid till fumes of sulphuric acid escape copiously. However, incomplete expulsion of fluorine, loss of phosphorus, forming of difficulty soluble anhydrous sulphates of aluminium, iron, chromium, nickel and calcium, as well as forming of insoluble phosphates when elements such as titanium and zirconium are present, and which may all occur if the evaporation of sulphuric acid is prolonged at high temperatures or carried to dryness, make decomposition by the use of hydrofluoric and sulphuric acids less attractive. Nevertheless in the usual case sulphuric acid must be used not only for the elimination of all fluorine but also to prevent loss of titanium by volatilization as fluoride.

The use of perchloric acid prevents the formation of an insoluble calcium salt but should not be used if an organic reagent such as alcohol or acetone has been used to assist in wetting the sample prior to the addition of hydrofluoric acid. The well-diluted perchloric acid has no hazardous properties, but the hot, concentrated, constant-boiling acid in contact with reducing matter, organic or inorganic, may produce violent and dangerous explosions.

Decomposition of some silicates may be effected by using nitric acid in connection with hydrofluoric acid, as with garnets. In this case 15 ml of nitric acid are added for each 25 ml of hydrofluoric acid. The mixture is evaporated to dryness on a steam bath and dissolved in a little water. If a residue is left, the procedure should be repeated. (Insoluble sulphates and phosphates are not decomposed by evaporation with nitric acid, and will be found in the residue if appreciable amounts of the reacting elements are present.) When no residue is left, all fluorine may be expelled by bringing the residue of salts in clear solution with dilute sulphuric acid and then evaporating a second time to incipient fumes of sulphuric acid.

When silicon is not to be determined, the following procedure is suitable to break up silicates using hydrofluoric acid.

Transfer 0.05–0.5 g of air-dry sample (–200 mesh to the linear inch) to a 25–100-ml well-cleaned, well-ignited, and cool platinum basin equipped with a platinum cover (see p. 6). Add 10–25 ml of hydrofluoric acid together with sulphuric acid or (and) one of the other acids, the choice and the amount of acid being governed by the composition and the amount of the material. Cover the basin with the cover and digest on the steam bath for 30 min with occasional and careful stirring with a platinum rod. Remove the cover, place the basin over a low flame, and evaporate under a good hood to fumes of acid. Cool, carefully wash the inside surface of the basin with H_2SO_4 (1 + 9), bring the deposited salts into solution as much as possible, evaporate again to fumes of acid and cool. (Both evaporations should not be continued to complete expulsion of the sulphuric acid, for the deposited salts may be hard to dissolve; moreover, elements such as titanium will hydrolyse easily if the solution is hot and its acidity is low.) Add 50 ml of water and heat on the steam bath until the soluble salts are in solution. Nature and amounts of the nonvolatile compounds left after the second evaporation and free from fluorine, can be determined by the methods of separation and determination mentioned at various places throughout the book (see also Scheme II, p. 105). If the solution is entirely clear, cool, and transfer to a suitable vessel or graduated flask, depending of the succeeding operations.

(Decomposition of the sample by direct treatment with hydrofluoric acid alone, or in combination with one or two of the other acids enumerated (p. 35), is more attractive than a fusion procedure because the introduction of foreign salts is avoided and silicon is eliminated. Silicates which do not respond to a wet attack, as well as those in which silicon is to be determined, must be broken up by a fusion method.)

Acid-resistant minerals and insoluble residues

With silicates which are for the most part decomposable by direct treatment with acids, it happens frequently that a small amount of minerals remain which resist attack by acids. These minerals usually consist of small dense black particles distinguished from particles of carbon or graphite by their readiness to sink rapidly to the bottom of the basin when the solution is agitated. These acid-resistant minerals must not be ignored as they may contain a fairly high proportion of some of the elements normally determined. The proportion of TiO_2, Cr_2O_3 and MnO, contained in the acid-resistant minerals from a selection of rocks, has shown that in some cases, e.g., in peridotites, dunites and eclogites, a large proportion of a particular element may be present. Because of this, acid-resistant materials or residues that remain undissolved after decomposition by acids, must be separated, fused with a suitable flux, the melts dissolved in water or acid, and the solutions so obtained added to the main solutions, if the solution of the melt has not to be examined separately.

PRECIPITATION

For the conversion of soluble into insoluble ions, precipitation is used in a large number of separations in silicate analysis. While some conversions are comparatively simple, others require strict control of the conditions under which the precipitations are made.

It is preferable to use small slips of pH indicator paper or some drops of an indicator solution rather than a pH meter where a check on the pH of the solution is required during precipitation. Quantitative cleaning of the electrodes presents a problem where the precipitate is not easily washed off by a stream of water, e.g., that of the ammonium hydroxide group. If indicator papers are to be used, they should be kept in a vapour-proof container to retain their sensitivity, while indicator solutions must be kept in the dark in bottles made of resistant glass.

Details of each particular precipitation are given in later chapters with the appropriate method but attention is drawn here to a few special subjects being discussed in the next six sections, viz.: co-precipitation, post-precipitation, peptisation, filtration, the washing of precipitates, and drying and ignition of precipitates.

CO-PRECIPITATION

Co-precipitation is the name given to the contamination of a precipitate by foreign ions present in the solution from which the precipitate is formed. The purity of a precipitate depends first upon the type of precipitate, i.e., crystalline or flocculent, and secondly on the nature and the amount of the foreign ions present in the solution in which the precipitate is formed. Foreign ions precipitate either in the structure of a crystalline precipitate (occlusion; almost synonymous with adsorption but of a chemical nature rather than due to physical forces, is directly proportional to the concentration of the undissociated foreign ions in the solution during precipitation) or are adsorbed as in the case of a flocculent precipitate (adsorption). The types commonly met in silicate analysis are shown in Table III.

TABLE III

Grades of Whatman filter paper required for various precipitates

Grade	*Coarse (41,541)*	*Fine (40,540)*	*Finest (42,542)*
Precipitates	silica (1st ppt) ammonium hydroxide group	silica (2nd ppt) calcium sulphate calcium oxalate	magnesium ammonium phosphate barium sulphate manganese dioxide

Crystalline precipitates are able to recrystallize upon standing. In the case of calcium oxalate and barium sulphate this is best achieved by standing on a hot steam bath for several hours, a proces termed "ageing". The particle size of the precipitate is increased by crystallization of material still in solution. There is some evidence of re-solution of impure crystals and a recrystallization of a purer form as being the mechanism by which "ageing" purifies a precipitate. However, if the impurity is isomorphous with the crystalline precipitate then "ageing" will not help since the impurity will be present in the lattice of the crystals.

The use of the flocculent ferric and aluminium hydroxides as a collector precipitate for traces of other elements is well known. Because of this property, it is seldom possible to avoid contamination of the ammonium hydroxide group to some extent by other elements. However, precipitation of the ammonium hydroxide group in a 2.5% solution of either ammonium chloride or ammonium nitrate (volatile electrolytes) under special conditions prevents precipitation of elements such as nickel, manganese, calcium and magnesium, limits the alkalinity of the solution, and aids in coagulation of the ammonium hydroxide group.

POST-PRECIPITATION

If, after the precipitation has been completed, extraneous elements having a common anion are present in the solution, these are able to crystallize on the precipitate causing contamination of the precipitate. This form of contamination is known as post-precipitation. An example of this is the post-precipitation of magnesium as magnesium oxalate on a calcium oxalate precipitate. A complication in this case is that for complete precipitation of the calcium in the presence of salts it is necessary to stand the solution overnight thus encourageing the post-precipitation of magnesium especially if it is present in quantity in the solution. It is therefore necessary to re-precipitate the calcium oxalate, in which case the precipitate will be surrounded by a much lower concentration of magnesium in the solution during this second precipitation. In the absence of a large quantity of salts, the calcium oxalate precipitates and "ages" in less time e.g., 5–6 hours, resulting in less post-precipitation of magnesium.

PEPTISATION

If adsorbed electrolyte ions are removed by washing a precipitate such as aluminium hydroxide (in the gel form of colloid) with distilled water, the electrolyte concentration may fall below the coagulation value and the precipitate (changed by this into the sol state of colloid) is then able to pass through ordinary filtering mediums. It is for this reason that acids or salt solutions (electrolytes) are used for the washing of precipitates. The appropriate washing solutions required for various gelatinous precipitates will be mentioned in the directions of separation and determination given in this book. After filtration the filtrate and washings should be allowed to stand in a vessel (of platinum, good porcelain, or Pyrex glass) for several hours to ascertain that no appreciable amounts of the precipitate have gone into colloidal suspension.

FILTRATION

Speed and accuracy in filtration depend to a large extent on the choice of proper filtration procedures and on the analyst's skill for carrying them out. Most filtrations are made through paper; grade and size of the filter paper should be chosen according to the nature of the precipitate and the use that is to be made of the precipitate or of the filtrate. If the precipitate and paper ash are to be weighed, the size of the filter (of known low ash content) should be adjusted to the amount of the precipitate; the precipitate should not occupy more than one third of its capacity.

Filter paper properly set in the funnel should reach to 5–15 mm below the rim, should fit close to the top and not touch the glass at the lower half.

Poorly made funnels, filters of the wrong texture or badly fitted, may cause the filtration not to get its full amount of washings; in some cases the long-lasting operation may also lead to changes in the precipitate (e.g., by oxidation by air). In general, a rapid filter should be used for gelatinous precipitates, a medium-texture paper for crystalline precipitates, and a close-texture paper for finest precipitates and precise work and for use with Büchner funnel and vacuum pump.

Macerated filter paper of known low ash content, like Whatman's "ashless" filter paper tablets, assists in the final filtration of a gelatinous precipitate such as that of the ammonium hydroxide group; added in a known quantity after the precipitate has been formed and then mixed with it, the precipitate is easily retained by the filter paper, the ignited residue will be finely divided and thus easily oxidable by continuing the ignition under good oxidizing conditions.

Fritted-glass filtering crucibles and -funnels of chemical-resistant glass, or porous-bottom porcelain filtering crucibles, of sufficient fine porosity, are recommended when the precipitate is only to be dried at low temperatures before weighing, or the use of strong acids or alkalies or the presence of oxidizing agents, render paper unsuitable (see p. 5); below 150°C crucibles and funnels made of glass are to be used, above 150°C crucibles of porcelain. In such cases, Gooch crucibles of different capacities with fixed or removable bottoms made of platinum, porcelain, or fused silica, may also be used by which the filter pad (weighing no more than 50–100 mg) may consist of three layers of different sizes (coarse, medium, and fine) of purified asbestos fiber of the amphibole variety resistant to chemical attack (see p. 12), or of a mat of pliable Pyrex glass wool, depending on the procedure to be followed.

In filtrations carried out by means of the filtering crucibles or funnels mentioned above, suction is required and the use of a thick-walled bell glass recommended enclosing the recipient and being strong enough to sustain a certain vacuum; provided with a ground-glass plate at the bottom and tubules at the top and side of the bell glass, as well as a two-way stopcock on the exhaust, the filtrate can be caught in the proper vessel in which it has to be treated.

In some cases filtration of large volumes of solution or the washing of certain precipitates by decantation, may be simplified by means of a centrifuge. For instance in the double precipitation of calcium as oxalate, it will be necessary to filter two portions of 250 ml of liquid through a small close-texture ashless paper and to wash the oxalate precipitate with small portions of ammonium oxalate (see p. 141). After the double precipitation of the oxalate group, time will be saved by centrifuging portions of the supernatant liquid in a waisted centrifuge tube and by washing of the precipitate by decantation, omitting to decant the last 1–2 ml each time. The decanted liquid and washings shall be collected in a clean beaker, carefully inspected for any precipitate brought over from the decantations, and the washed oxalate precipitate reserved for the determination of calcium.

WASHING OF PRECIPITATES

The choice of the washing solution, a rough estimate of the volume and the number of washings to be applied in every case, as well as testing of the filtrate and washings to ascertain whether all of the objectionable matter is removed, are important factors in the accuracy of a determination. For instance in a direct gravimetric determination of silicon and magnesium, insufficient washing will cause a positive error by respectively sodium chloride contamination of silica, and ammonium phosphate contamination of magnesium pyrophosphate, and a negative error in a direct gravimetric determination of iron by incomplete removal of ammonium chloride from the ammonium hydroxide group due to volatilization of ferric chloride during ignition. On the other hand, excessive washing will tend to dissolve some of the precipitate to be determined, the reason why the volume of the washing solution should always be held down to the smallest amount required to remove the objectionable matter.

No general rules can be given for the washing of precipitates, because the way to wash depends on the character of the precipitate under test. For instance, stable precipitates somewhat soluble should be washed with a number of small portions of washing solution that are well drained instead of with large portions, while precipitates that tend to clog the filter or form channels or oxidize quickly when allowed to drain, should be covered with washing solution and filtered by decantation to remove the objectionable matter. However, in most cases the washing in the filter should be started by directing the jet of washing solution at the clean upper part of the filter to prevent loss of precipitate by ejection, after which the jet should be lowered gradually for washing the precipitate and working it down into the bottom of the filter. If filter and precipitate have to be washed thoroughly, and the precipitate tends to creep (like e.g., freshly precipitated calcium oxalate and magnesium ammonium phosphate), special care must be taken in fitting the filter and in keeping the wash solution about 1 cm from the rim to prevent loss of precipitate; after washing any precipitate adhering to the sides of the funnel should be gathered on a moisted slip of filter paper which is then added to the filter containing the bulk of the washed precipitate. Tests for the completeness of washing should always be made in small portions of the washing solution drained through the filter.

DRYING AND IGNITION OF PRECIPITATES

No general rules can be given for the drying and ignition, for the treatments depend on the character of the precipitates; such treatments may vary from gently drying to strong ignition at 1,200°C (in air, oxygen, hydrogen or nitrogen) in vessels made of proper material.

If the precipitate can be safely ignited in contact with charring filter paper (viz., when the reduced material is nonvolatile and reoxidable after the carbon has been burned off, also does not attack the vessel) as in the majority of cases, the moist paper can be wrapped about its contents, the whole placed in a crucible made of proper material, and dried, charred and ignited as follows.

Fasten an iron support ring to an iron support, place crucible with contents, tilted at about 45° in a platinum or silica triangle on the support ring, place a Bunsen burner (fitted with a star-support and cone) with a low flame protected from draughts about 2″ below the rim of the crucible and heat carefully without spattering or boiling inside of the crucible until dry. Then raise the flame a little so that after a few minutes vapors gently escape from the crucible and carbonization of the paper takes place. (Fast carbonization or inflaming of the paper, causing reduction or mechanical loss of precipitate, should be avoided!) When vapors no longer come off, direct the flame against the bottom of the crucible, and burn the carbon off slowly raising the flame when necessary, until the carbon has fully disappeared. Cover the crucible with a well-fitting lid, increase the flame to its maximum and if the full heat of a Bunsen flame is insufficient to attain the temperature required, make a change over to a Meker burner or electric muffle, heating according to the directions given below. Cool in a desiccator over a good desiccant to room temperature, weigh while covered, and repeat the ignition and weighing until constant weight is obtained; if necessary, correct for the loss of weight of the platinum crucible. The weight of the ash of good quality filter paper may be neglected.

Directions

(*1*) The inner blue cone of a nonluminous flame from a Bunsen burner, and the green zone of a Meker burner, should not touch a platinum vessel to prevent damage of the vessel resulting from formation of platinum carbide.

(*2*) The flame should not envelop the crucible fully; if it does, the air within the crucible will be replaced by a mixture of reducing combustion products and water, resulting from incompletely burned gases causing the reduction of reducible substances like ferric oxide or preventing complete dehydration at normal temperatures because of the presence of water vapor. To exclude conversion by flame gases, the crucible with contents may be placed upon a shield of stiff platinum foil in which a circular opening has been bored large enough to admit the crucible to two thirds of its height. This shield is laid over an asbestos board of 5″ sq. in which a circular opening of much larger diameter than that of the shield has been made, after which the board should be fastened to an iron support and given the inclined position required, so as to allow the flame gases to envelop the bottom of the crucible only.

(*3*) The final heating of precipitates that are reducible by charring paper or tend to lose oxygen on very strong heating, should be done as to insure access of air to the interior of the crucible, whereas at the end of the ignition the lid should be lifted for a moment to prevent any gases (water vapor, carbon dioxide, etc.) that are left in the crucible being absorbed by the residue as it cools in the desiccator.

(*4*) For ignitions as in (*3*) and for those involving very prolonged heating an electric muffle (preferably equipped with a thermocouple and indicating pyrometer) is recommended on account of its non-reducing atmosphere.

(*5*) If the substance does not attack platinum, for ignitions at high temperatures by maximum flame, crucibles of platinum are to be preferred to those made of quartz or porcelain because of the higher temperatures obtainable in their interiors; no mutual differences in temperature appear when using muffles.

(*6*) When heated above 1,000°C slow variable losses in weight of platinum vessels may occur due to the volatilization, particularly of indium when alloyed with the platinum; the losses increase with the iridium content of the vessel, the temperature and the length of heating. The rate of loss in weight of the empty vessels should be ascertained previously so that a correction may be applied when very long ignitions at high temperatures should be made, or else the vessel should be reweighed after removing the ignited residue avoiding attack on the vessel.

(*7*) In the usual case, an ignited platinum crucible and residue should be cooled in a desiccator for 20 min, a silica or porcelain crucible and residue for 30 min. Ignited moderately hygroscopic residues should be cooled and weighed in crucible with well-fitting lids, whereas after reignition for a further 10 min reweighings should be made with the previous weights already placed on the balance so that the weight can be quickly obtained.

(*8*) For the weighing of ignited excessively hygroscopic residues that are to be cooled and weighed with minimum contact of air, weighing bottles of Pyrex glass which have two small corresponding holes (viz., one into the cover and one into the bottom) will be of great use. Then, without disturbing the lid, the hot crucible containing the ignited residue is placed in the weighing bottle, the cover is put on and turned so that the two small holes will form the smallest possible hole to permit equalization of pressure. The whole is transferred to a desiccator, cooled, and weighed against objects of similar size and material, treated in a similar way, as tares.

CHAPTER 6

Chemical Analysis of Silicate Rocks

THE CONSTITUENTS OCCURRING IN ROCKS

Minerals, being homogeneous substances of a definite inorganic chemical composition and molecular structure, exhibit wide variations in the number and percentage of their elemental constituents of which one or more may be present as predominating constituents. Rocks are mixtures of different minerals, each having characters of its own; the elements most often found in rocks are: Si, Fe, Al, Ca, Mg, Na, K, H, and O, as major constituents; Ti, P, and Mn, as common minor constituents; and Cu, Cr, V, Zr, Ni, Co, Zn, Sr, Ba, Li, S, C, F, Cl, and N, as minor constituents that may be present in determinable or readily discoverable amounts. Seldom or mostly in insignificant amounts are encountered: Cb, Ta, W, Pb, Sn, Mo, Ag, Au, Pt, Be, U, Th, the rare earths, B, and He.

The general chemical character and the relations of the various constituents for the close classification of a rock may be found by following the arrangement proposed by Washington (1900), viz.: SiO_2, Al_2O_3, Fe_2O_3, FeO, MgO, CaO, Na_2O, K_2O, $H_2O + 110°C$, $H_2O - 110°C$, CO_2, TiO_2, ZrO_2, P_2O_5, SO_3, Cl, F, S (usually as pyrite), Cr_2O_3, V_2O_3, NiO, CoO, ZnO, CuO, MnO, SrO, BaO, Li_2O, C (usually as graphite), and NH_3, of which constituents the percentage figures and the molecular ratios are to be determined and calculated. For purposes of scientific interpretation and comparison the analyses should be made as complete as possible; in that case to the foregoing list might be added the constituents of uncommon occurence mentioned above of which some occur at times in determinable amounts. Though desirable in special cases, ordinarily it will not be necessary to subject silicate rocks to very complete analyses because a knowledge of the quantitative relations of those constituents that can be determined or readily discovered in a given amount of the sample (e.g., 1.0 g for SiO_2, Al_2O_3, etc.; 2 g or more for certain constituents; 5 g or more for CO_2) will suffice in the majority of cases. Whenever possible, a microscopic and spectroscopic examination of the rock should preceed the chemical analysis to detect the presence of extraneous constituents, or of more than commonly present quantities of certain constituents giving rise to difficulties in the analysis, in which cases the existing analytical procedure should be altered.

In view of the elements encountered in rocks, those constituents are chosen that can be determined conveniently in one and the same portion of the sample. Consequently, it is usual in silicate analysis to determine in the main portion (1.0 g): SiO_2, total iron, TiO_2, MnO, NiO, CaO, SrO, MgO, and the weight of the ignited ammonium hydroxide precipitate consisting of Al_2O_3, all iron as Fe_2O_3, TiO_2, P_2O_5, ZrO_2, Cr_2O_3, practically all vanadium as V_2O_5, and rare earths, if they are present in the rock; in another portion (0.5–1.0 g): FeO, and the total iron (the last to be used as check); in a third portion (2.0 g): BaO, ZrO_2, rare earths, and the total sulphur; in a fourth portion (>2.0 g): V_2O_3, and Cr_2O_3; in a fifth portion (0.5 g): the alkalies; in other portions: any other constituents by a proper combination of methods.

Because the composition of rocks is limited to varying percentages of a comparatively few major elements, a general procedure for the analysis of rocks has been developed to obtain first of all an idea as to the character of the rock under test; this "General Procedure" refers to the scheme of analysis in which the main portion taken from the sample is decomposed, groups of elements that exhibit a common reaction are first separated, and then the groups are examined for the members that may be present, and their amounts determined by suitable methods.

The General procedure almost universally followed, separates the elements into six groups: Acid group, Hydrogen sulphide group, Ammonium hydroxide group, Ammonium sulphide group, Ammonium oxalate group, and Ammonium phosphate group, which groups are obtained by respectively decomposition of the sample and evaporation of its solution with hydrochloric acid, and successive precipitations with hydrogen sulphide, ammonium hydroxide, ammonium oxalate and diammonium phosphate, as detailed below.

THE GENERAL PROCEDURE
(The separation of the six groups of elements)

Acid group

In silicate analysis, silicon being the major acid-forming constituent of the Acid group, the colloidal silicic acid formed by fusion of the sample with a suitable flux followed by solution of the melt in one of the common acids, is dehydrated and thus rendered insoluble by evaporating the solution to dryness. Hydrated silicic acid is appreciable soluble in acid solutions, therefore the acidity of the solution in which the melt is digested as well as the period of digestion should be no greater than is necessary for the solution of accompanying salts and the prevention of hydrolysis. When double evaporations with hydrochloric acid are to be made, the main lot of silicic acid should be washed with hot HCl (5 + 95) until the colour of the filter paper indicates removal of most of the iron; hot water may be used for the last washings and the filter paper should be sucked dry at the pump. The silicic acid residue of the second filtration should be washed with sufficient cool HCl (1 + 99) and after that with hot water. The foreign ions, the removal of which is necessary, are thus leached from the silicic acid particles giving a small amount of impurities after ignition and treatment with HF-H_2SO_4. The smaller the amount of impurities in the combined silica residues, the smaller the error associated with weighing the impurities as sulphate instead of oxide; however, the non-volatile matter should always be ignited at a temperature high enough to change it to the state in which it occurred in the impure silica.

As mentioned above, the separation of silicon is never complete because of the solubility of silicic acid in acid solution. When the solubility equilibrium is reached, further dehydrations and filtrations are ineffective. The small amounts of silicic acid remaining in the filtrates after two evaporations, may be recovered and determined as silica in the ignited precipitate produced by ammonium hydroxide if a fair-sized iron or aluminium hydroxide precipitate is obtained, whereas in case of doubt or if very accurate results are desired, the full amount of residual silicon that may be still in solution can be estimated by colorimetric procedures and is reported as silica (p. 78). According to the General procedure the filtrates or solutions left after the separation of the Acid group and the determination of the silica are reserved for further investigations (see Scheme I, p. 85).

Hydrogen sulphide group

In silicate rocks appreciable amounts, if any, of the members of the Hydrogen sulphide group (like copper, cadmium, lead) can hardly be expected. Moreover, the reserved filtrates or solutions left after the determination of silica (see Acid group, p. 49), are too acid for the complete precipitation of the sulphides of the minute amounts of the minor hydrogen sulphide elements; if present, these elements are to be detected and determined by special methods. Hence in silicate analysis passing hydrogen sulphide through the acid filtrate from the silicic acid may serve only for the removal of the platinum metals introduced through fusion or evaporation in platinum ware, their filtrate after expelling the hydrogen sulphide and reoxidation of the iron to the trivalent state, being reserved for the precipitation of the Ammonium hydroxide group (p. 51). Causing no difficulties, however, when present alone, the platinum metals can be removed more easily at a latter stage by other methods.

Ammonium hydroxide group

Most ammonium hydroxide precipitates are voluminous owing to their colloid nature. For the reason that colloids in the sol state pass through ordinary filtering mediums, the presence of an electrolyte is necessary to prevent precipitation of elements such as magnesium, to limit the alkalinity of the solution, and to aid in coagulating the precipitate. Besides this, a washing solution containing a volatile electrolyte such as ammonium chloride or nitrate must be employed preferably similar to the solution in which the separation was made; the former to be preferred if only little iron is present and the filtrates and washings are to be acidified and evaporated in platinum, the latter in the final washing if much iron is present and subsequent operations in the filtrate permit.

In silicate analysis the precipitation of the members of the Ammonium hydroxide group (preferably associated with all the residual silicon, phosphorus, and vanadium, if present in the proper ratio to the precipitable elements), demands careful attention to the hydrogen ion concentration of the hot solution, a rapid filtration using a medium-texture ashless paper, and washing of the precipitate with hot ammonium chloride (2%), with a single final washing with about 5 ml of hot water if much iron is present; the paper should be filled no more than one half with precipitate, the solution should be kept about 1 cm from the edge, the precipitate should be kept wet because it forms channels when allowed to drain, and most of the washing should be done by decantation. Resolution and reprecipitation in the original beaker, and filtration of the Ammonium hydroxide group adding ashless paper pulp after reprecipitation, may be necessary if nature or amounts of other elements call for it. The filtrates and washings from the ammonium hydroxide precipitate are reserved for the determination of the other constituents; at this point the General procedure should be varied, depending on whether or not an ammonium sulphide separation is necessary.

If the sample contains amounts of copper, nickel, cobalt, or zinc, that can be ignored, the use of ammonium sulphide can be omitted.

In that case, all members of the Ammonium sulphide and copper groups, if present, together with the introduced platinum metals, will be found (wholly or in part) in one or both of the two succeeding groups, the amount of each found in the Ammonium oxalate group always being much smaller than that found in the Ammonium phosphate group.

When the sample contains the elements mentioned above in determinable or disturbing amounts, an ammonium sulphide separation is required and carried out in accordance with the directions given under Ammonium sulphide group.

Ammonium sulphide group

After the separations of the Acid and Ammonium hydroxide groups (pp. 49 and 51), copper, nickel, cobalt, and zinc, if present in determinable amounts, are completely, manganese (together with a small part of the introduced platinum metals) incompletely, separated from the alkaline earth and magnesium by rendering the combined filtrates and washings left after the separation of the Ammonium hydroxide group neutral with ammonium hydroxide, treating with colourless ammonium sulphide and allowing the solution to stand overnight. The solution is then filtered uninterruptedly and the sulphides are washed with ammonium chloride (2%) containing a little colourless ammonium sulphide. The precipitate may be kept for the identification (seldom for the quantitative determination) of copper, nickel, cobalt, or zinc, whereas the filtrate and washings are reserved for the determination of the other constituents.

Ammonium oxalate group

After removal of the Acid, Ammonium hydroxide, and Ammonium sulphide groups, the constituents that may remain in solution are the alkaline earths, magnesium, the alkalies, and a small part (usually not over 0.2 mg) of manganese that escapes precipitation as sulphide. In the absence of phosphate or sulphate ions, and preferably present as chlorides in hydrochloric acid solution, the separation of the members of the Ammonium oxalate group (viz. calcium and strontium) from those of the succeeding Ammonium phosphate group (barium and magnesium) by ammonium oxalate, is influenced by their relative proportions. However, in most silicate rocks the percentages of strontium and barium found are so small (usually below 0.3% and 0.1% respectively) that in general calcium and magnesium may be considered as to be present alone; calcium being preponderant, may be separated from magnesium and the alkalies in alkaline solution by double precipitation with ammonium oxalate, and washed with ammonia oxalate (0.1%) as detailed on p. 141.

The precipitation of calcium as oxalate is never complete; a small amount (usually not over 0.5 mg) of calcium will escape double precipitation in alkaline solution with ammonium oxalate and be precipitated as phosphate along with the Ammonium phosphate group, thus causing a minus error in the determination of calcium oxide, and a plus error in the subsequent determination of magnesia. However, in precipitating calcium in alkaline solution with oxalate, magnesium is occluded and adsorbed to some extent by the calcium oxalate causing a plus error in the determination of calcium oxide, and a minus error in the determination of magnesia. Because as a rule both errors tend to compensate if double precipitation and washing have been carried out under proper conditions, in ordinary analyses no attempt is made to correct for the co-precipitated element in question. The precipitate is kept for the determination of calcium (and of strontium, if any may be present), whereas the filtrate and washings are combined with those of the first precipitation and reserved for the precipitation of the Ammonium phosphate group (p. 53).

When the sample contains elements such as manganese, cobalt, and nickel in appreciable amounts, and a preliminary treatment with ammonium sulphide has been omitted, the minor part of these elements will be caught with the Ammonium oxalate, the major part with the succeeding Ammonium phosphate group.

Ammonium phosphate group

In the General procedure the precipitation of the members of the Ammonium phosphate group (viz. barium and magnesium) follows preliminary separations of the members of the Acid, Hydrogen sulphide, Ammonium hydroxide, Ammonium sulphide, and Ammonium oxalate group. Precipitation with hydrogen sulphide can be omitted if elements such as copper, cadmium, or lead, are present only in small amounts. Precipitation with ammonium sulphide can be omitted if elements such as manganese, copper, nickel, cobalt, or zinc, are absent. In most silicate rocks manganese is the only one of the five that may be present in disturbing amounts; if ignored in an ordinary analysis, the minor part of it being caught with the Ammonium oxalate group. The major part of the element, together with magnesium, will be precipitated by diammonium phosphate in ammoniacal solution, determined by colorimetry, and its effect, as $Mn_2P_2O_7$, subtracted from the ignited and weighed ammonium phosphate precipitate.

Besides manganese, the weighed residue can be expected to contain all of any calcium and traces of strontium that escaped precipitation as oxalate, and practically all of the barium, if these elements were present in the original sample; barium only if no sulphates were present or introduced during the analysis. However, the amounts of strontium and barium usually encountered in silicate rocks are so small (see p. 47) that in general one has to deal with a filtrate from the Ammonium oxalate group of the sample, containing practically all of the magnesium, the major part of the manganese, and a small amount (usually not over 0.5 mg) of calcium which escaped precipitation as oxalate, besides more or less accumulated ammonium and alkali salts resulting from prior fusions or separations.

If very little magnesium is present (as e.g., in silicates of the feldspar group) and prior operations have introduced any considerable amount of ammonium salts (retarding the precipitation of magnesium and barium), the accumulated ammonium salts in the combined filtrates from the Ammonium oxalate group must be destroyed by wet attack, the dry residue dissolved in an appropriate quantity of dilute hydrochloric acid, the solution made slightly ammoniacal, filtered, the washed impurities discarded, and the filtrate diluted to a convenient volume before the first precipitation by diammonium phosphate must be made.

When appreciable amounts of magnesium are present in a moderate volume of solution and undue amounts of alkali salts (particularly potassium salts) are absent, a preliminary evaporation of the filtrate from the Ammonium oxalate group or removal of alkali salts from it is unnecessary when double precipitations with diammonium phosphate are made; in that case the impure precipitate first obtained and dissolved in dilute hydrochloric acid may be reprecipitated under favourable conditions without significant loss of magnesium. Precipitations are made in a solution containing 5–10% v/v of ammonium hydroxide, the precipitates allowed to stand for some time, washed with NH_4OH (5 + 95), and the filtrates and washings discarded. The second precipitate of magnesium thus obtained is gradually heated to about 1,100°C, weighed as $Mg_2P_2O_7$, and tested for the more common contaminants (barium, calcium, and manganese) as described under Magnesium (p. 153). Under such conditions the precipitation of calcium and manganese that may have escaped precipitation by ammonium oxalate, is complete; that of barium is not complete because small amounts of the element escape double precipitations with diammonium phosphate. If present in determinable small amounts, corrections for barium and calcium in the ignited and weighed ammonium phosphate precipitate are made on the basis of the normal phosphates, $Ba_3(PO_4)_2$ and $Ca_3(PO_4)_2$, those for manganese on the basis of the pyrophosphate, $Mn_2P_2O_7$.

Alkali group

If originally present in the sample and double precipitations are made, the combined filtrates and washings from the Ammonium phosphate group (p. 54) obtained according to the General procedure, can be expected to contain all of the alkalies (sodium, potassium, lithium, rubidium, and cesium), traces to small amounts of a few less common elements (like nickel, strontium, and barium) through incomplete precipitation in their own groups, and sulphur when originally present as sulphate unless associated with elements such as barium and strontium; in addition, the excess of phosphate added for the precipitation of the Ammonium phosphate group besides more or less accumulated ammonium and alkali salts and platinum resulting from prior fusions or separations. Thus, the foregoing separations do not leave a combined final filtrate that can be used in determinations of the remaining constituents and the actual determination of the alkalies should be made as described in Chapter 17, (flame photometry) in a separate portion of the sample.

CHAPTER 7

Preparation of the Laboratory Sample

GENERAL REMARKS

An accurate analysis will be only possible if the sample is both representative of the rock or mineral and as free as possible from contamination of drills, milling cutters, crushers, grinders, mixers, mortar, sieves, etc., which may cause trouble and must be taken into account. Although free from contamination of major elements it must be remembered that at a later date the sample may be required for use in trace element analysis. Because of this, apparatus for the preparation of analytical samples should be strongly built and the crushing plates should be made of a hard and abrasion-resistant steel, dividers should be of the enclosed type to reduce dust losses, mortar and pestle should both be made of properly hardened alloy steel of a kind and grade designed to resist severe abrasive forces. Sieves should be made of the best silk bolting cloth, because appreciable contamination may arise from the use of brass or other metal sieves in relation to trace elements. St.Martin bolting cloth (Henry Simon, Ltd., Cheadle Heath, Stockport, England) is satisfactory for this purpose; a separate piece should be cut for the sieving of each sample. Either wooden or plastic rings are used to hold the cloth taut while sieving. The surface of the cloth should not be brushed, or any object placed on it, with the purpose of hastening the passage of particles. Any damage to the cloth will result in larger than required particles passing through the mesh. Material adhering to the surface of the sieve should be removed by inverting the sieve and tapping gently on the retaining ring.

Before preparing the analytical samples all equipment should be carefully cleaned to prevent introduction of undesirable contaminants into the samples. The roughened surfaces of the apparatus such as pounders, steel plates and jaw pinchers are frequently pitted owing to impact with particularly hard minerals. These pits are especially liable to contain particles of previously crushed samples and are best cleaned by passing a hard steel brush over the surfaces from several directions. The loose material is then either blown off with compressed air or shaken off by inverting the surfaces. An agate mortar, when used in the final grinding, should first be cleaned with nitric acid and rinsed with water, a sufficient amount of wet "acid treated" quartz sand should be ground in the mortar to a fine powder rubbing the entire surface of the mortar, and finally mortar and pestle rinsed with water and dried.

The original size of the specimen will depend largely upon the size of the constituent minerals. Pegmatitic rocks will require a larger specimen than fine grained rocks for representing their true average composition. A detailed discussion of the sampling of various geological materials has been given by Shaw (1961) and Wilson (1964).

Crushing

After the large sample is received, it is first crushed to lumps of reasonable sizes by means of a jaw pincher (Fig.3) or by a specially hardened steel plate and pounder (14 lb.); using the two latter, a wooden fillet 2–3 inches in height surrounding the steel plate will assist in restraining the flying chips. Another piece of equipment often used at this stage is a specially hardened steel mortar and pestle, by virtue of its construction, obviating loss of material by flying fragments.

Great care must be exercised if the sample is to be truly representative. If the material is fine grained, very fine dust-like material should be discarded and only the chips used for the next stage in the reduction. If the material is coarse grained (and even uniform in structure), the different-sized particles obtained by crushing may differ in composition because the more brittle constituents will con-

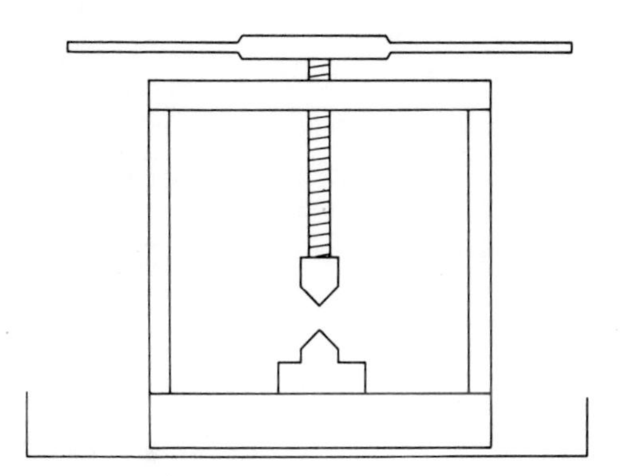

Fig.3. Apparatus for pinching off portions of rock.

centrate in the finer portions; in that case the dust-like material must be collected and added to the chips.

According to the general procedure for reducing samples of ore to laboratory size, the large sample is crushed to 1/8 inch size, the well-mixed sample formed into a flattened pile by a spiral motion of the shovel, the pile divided into four quarters by drawing lines at right angles through the center, two diagonally opposite quarters are removed and the remaining quarters thoroughly mixed. This operation is repeated until a laboratory sample of about 1 kg is obtained (Fig.4) after which the sample is transferred in small portions to a surface-hardened tool steel percussion mortar (Fig.5). As the name implies, the reduction in size is effected by percussion and not by grinding; serious contamination will result even from the least grinding motion with the piston.

When a portion of the laboratory sample has been transferred to the mortar, the piston is replaced and the collar held down firmly with the left hand while the head of the piston is struck by a vertical blow of a 1/2 lb. hammer. Holding the collar close to the base plate will reduce the movement of the sample.

After the first blow the piston is rotated through 90° and struck again, etc. In most cases, after about six blows, sufficient of the

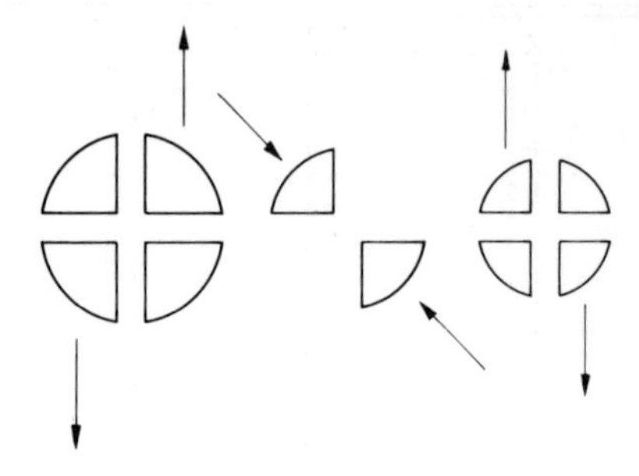

Fig.4. Cone and quartering procedure. The sample is thoroughly mixed on a clean sheet of paper using a horn of plastic spatula. The sample is then shaped into the form of a cone and divided into four equal portions. Opposite segments are removed from the sheet and the two remaining segments mixed again. Each time a division is made, opposite segments are discarded. This procedure is repeated until the quantity remaining is that required for the final sample. It is essential when using the technique that the sample is of an even grain size.

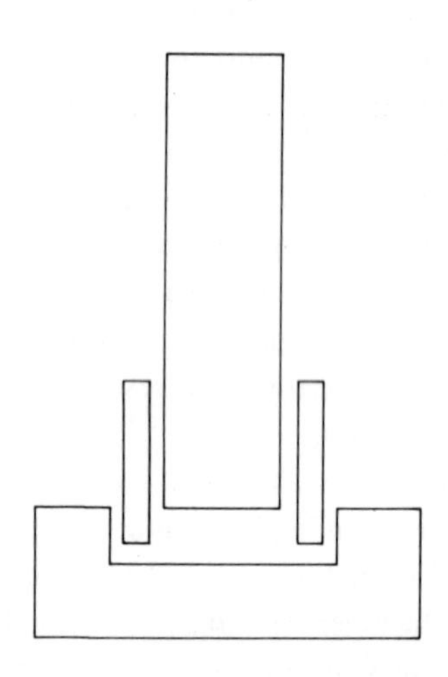

Fig.5. Percussion mortar used for crushing samples. The surfaces used for crushing are usually faced with tungsten carbide to avoid excessive wear.

sample is reduced to –30 mesh to warrant its transfer to a cylindrical collecting glass about 4 cm high and 8 cm in diameter.

When the whole laboratory sample has been crushed and collected, a piece of best silk bolting cloth with 30 meshes to the linear centimeter is stretched over the open side of the glass and clamped fast by a wooden ring about 1 cm broad. The glass inverted, the powder is carefully sifted onto a large sheet of glazed paper, the remaining large particles are recrushed in the mortar and resifted until nearly all have passed the sieve, after which the whole final small portion of oversize particles being ground in an agate mortar must be added to the sifted powder. Applying this procedure, serious contamination by metallic iron is avoided, the introduction of bolting cloth will be negligible, and a fairly representative sample (for the most part being of desired even grain size) is obtained.

The sample –30 mesh to the linear inch and well-mixed to uniformity, is now reduced by quatering or by means of a splitter (Fig.6), an apparatus into which the sample is poured, half being saved while the other half is put aside for storage (Pouring the sample into the top of the splitter, care should be taken that segregation

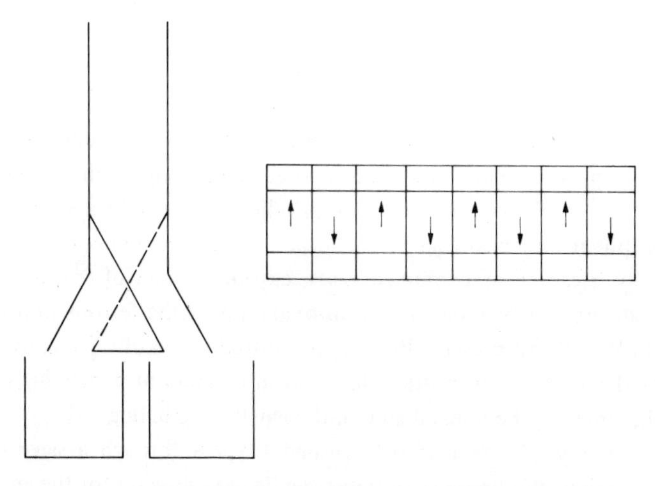

Fig.6. Sample splitter. The sample is spread evenly across the grill, arrows indicate the direction of flow of powder.

of different size particles and dust does not occur. A simple and inexpensive divider constructed from a turntable and a plastic container has been developed by Rowland, 1963).

The half of the sample that has been saved, is well mixed and passed once again through the apparatus, crushed if necessary to a somewhat smaller size and again mixed and divided into two or more fractions by the use of these or similar procedures, and the operations repeated until a test sample is obtained large enough for the analyses to be made. (The selection of correct portions of material and the preparation of a representative test sample from such portions are necessary prerequisites to every analysis, the analysis being of no value unless the test sample represents the true composition of the material from which it was selected.)

Although only 0.5–1 g may be required for the main portion (Scheme I, p. 85) the weight of the representative test sample should not be less than 10 g, and preferably 20 g, in case it should be advisable to use large quantities for certain determinations (notably CO_2), for standard material in developments of new methods, for trace element work or age determination.

Grinding

As a rule the analysis sample should be crushed and ground to the *coarsest* powder that can be decomposed completely by the method of attack employed; needless fine grinding is not only a waste of time, but it may also lead to chemical changes in the sample caused e.g., by escaping of contained gases, expelling water of crystallization absorption of moisture, or air-oxidation of the ferrous iron or sulphide material present. Besides this, introduction of silica derived from the mortar and pestle will occur in an amount depending on the hardness of the mineral and on the length of grinding.

Ordinarily silicate material crushed to pass through a sieve of 30 meshes to the linear centimeter can be used directly for the analysis, except for those portions in which ferrous iron and the alkalies are to be determined or constituents are to be removed by acids, which procedures require a powder of considerable fineness. In those cases the final grinding must be carried out on small quantities of the crushed −30 mesh material, with proper precautions that no chemical changes occur or selective dusting takes place.

Although by grinding in an agate mortar, introduction of silica into the sample may be considerable and sometimes may reach one half percent of the weight of the ground material, the quantity abraded is hardly of moment in comparison with the amount of silica invariably present in silicates. Therefore, notwithstanding the fact that by grinding hard materials like glass, using mortars and pestles of hardened alloy steel or agate, the abrasion of steel is about a tenth of that of agate, the choice in favour of one of the grinding implements should be made in regard to the least disturbing effect on the outcome of the determinations which are to be made.

MINERALS

Where minerals have been separated from the host rock by means of heavy liquids, the grains should be thoroughly air-dried under a cover placed a few centimeters above the mortar and pestle before crushing. In general, mineral samples will require a smaller sample for representing their true average composition. If the host rock has been crushed (which is usually the case) to release the minerals, the initial mesh size of the sample will be fairly small. When a portion of the sample is required for optical examination, this should be set aside before crushing is commenced.

"Hard minerals" (viz. minerals whose Mohs' hardness $\geqslant 7$, like tourmaline, garnet, and topaz; quartz = 7) must be reduced to a mesh size just fine enough for the attack that is contemplated, by crushing in a mortar with pestle both made of properly hardened alloy steel; the use of this steel implement does not affect the composition of the sample in contrast to that of agate by which the abrasion of mortar and pestle becomes more manifest and introduction of silica into the sample (in an amount depending on the hardness of the mineral and on the length of grinding) may easily exceed one percent of the weight of the same sample crushed in a hardened alloy steel percussion mortar.

Data obtained in the course of thousands of rocks and minerals have shown that: (*1*) As a rule silicate material crushed to pass a sieve of 30 meshes to the linear centimeter can be used directly for the analysis without further grinding in an agate mortar. (*2*) Needless fine grinding of the analysis samples may not lead to chemical changes in the samples only, but also to introduction of appreciable material from the grinding implements; see Grinding, p. 62. (*3*) For powders (containing ferrous iron) being completely decomposed within 20 min by boiling with dilute hydrofluoric acid, a determination of ferrous iron carried out with proper precautions will give results close to their true ferrous content. (*4*) Grinding in absolute alcohol (as nonoxidizing medium) is fairly satisfactory to prevent oxidation of ferrous iron as well as of other oxidizable constituents of the sample.

For those reasons, if test samples are to be prepared using the laboratory sample crushed to –30 mesh to the linear centimeter as starting point, a sieve size should be chosen that will afford the coarsest powder that can be decomposed completely or nearly completely within 20 min by boiling with dilute hydrofluoric acid in a platinum crucible.

If after treatment with hydrofluoric acid and evaporating to dryness too much undecomposed material may be left and grinding is required, to secure sufficient fineness and to avoid oxidation of the sample a weighed portion (0.5–1.0 g) of the laboratory sample –30 mesh should be ground under absolute alcohol (using small quantities at a time) in an agate mortar only long enough as to furnish a powder that will leave little or no residue when treated with hydrofluoric acid. If grinding under alcohol has been necessary, a pyrex cover glass (to prevent introduction of dust and allowing circulation of air) should be placed a few centimeters above mortar and pestle. All alcohol having volatilized spontaneously, the powder moistened with water should be transferred quantitatively by means of a fine jet of water to a suitable platinum crucible, cleaning and rinsing mortar and pestle carefully before treating the test sample with hydrofluoric acid.

Laboratory samples free from alteration products, crushed to pass a sieve of 30–60 meshes to the linear centimeter yielding relatively coarse test samples fit for analysis and mostly containing less than 0.1% of moisture, should be preserved in well-stoppered bottles to secure a uniform hygroscopic condition for comparison of analytical results.

If a very fine powder must be employed and a very fine state of division of the test samples is required (as with samples containing a considerable percentage of e.g., the minerals garnet, staurolite, topaz, andalusite, sillimanite, cyanite or chrysolite), attention should be paid to the influence of fine grinding on the water and ferrous content of rocks and minerals. The finer the material is ground and the longer the time of grinding and exposure to the air, the more atmospheric moisture may be taken up (partly retained at temperatures far above 100°C), the more water of crystallization may be expelled, and the less ferrous iron may be found.

With different minerals, different degrees of oxidation are brought about under like conditions; readily oxidizable by grinding in air are i.a., garnet and staurolite, less oxidizable are augite and hornblende. In such cases, fine grinding should always be done on a weighed small quantity of the crushed 30-mesh laboratory sample with the same precautions as described above; the powder obtained should be used directly for the analysis, all the separate portions necessary for the analysis being weighed one after the other to eliminate the error due to difference of hygroscopicity in dry and moist weather.

METEORITES

Silicates without metal, e.g., achondrites

Where metal is absent, the sample may be treated as normal silicate material except that the fusion crust should be removed before crushing the sample. If the material is hard, removal of the fusion crust will probably require the use of a jaw pincher but a pair of steel cutters may be sufficient, provided all parts of the equipments contacting the samples are of steel showing a high resistance to abrasion.

Silicates with metal, e.g., chondrites

This type of material makes it one of the most difficult to sample properly since silicate, sulphide, phosphide and metal phases are present in the specimen. Any fusion crust should be removed beforehand, after which the sample, pinched into chips (ca. 1–2 cm cubes), should be crushed, one by one, in a hardened alloy steel percussion mortar.

If the sample is to be analysed by separating the metallic part magnetically from the non-magnetic material (Prior, 1913), both being treated as separate portions, the sample should be crushed without loss by dusting or otherwise to the coarsest powder that is fit for effective magnetic separation. Since the particle size of the metallic fraction varies, no fixed mesh size can be given to which the particles should be reduced but 30 meshes to the linear centimeter would be suitable in most cases. The metallic fraction is removed using a magnetic comb, weighed and separately analysed. The non-magnetic fraction further reduced to –100 mesh to the linear inch (40 to the linear centimeter), is also weighed and analysed, and the two analyses are combined in the ratio of their relative weights.

If the sample is to be subjected to chlorination (Moss et al., 1961, 1967), the sample should be in the form of powder reduced to –100 mesh.

A plentiful supply of liquid air is necessary if the sample is to be completely reduced to –200 mesh to the linear inch and analysed as a single sample (Easton and Lovering, 1963). This crushing technique which utilizes the brittleness of metal when cooled to a very low temperature has been described (Berry and Rudowsky, 1966).

The percussion mortar is used initially to reduce the mesh size of the chips to ca. 100 mesh to the linear inch, taking care that the pestle is only lightly struck so that the metal particles of the sample are not flattened (Sieving at very frequent intervals to remove the finer particles will reduce embedding of silicate particles in the metal grains). The powder that has passed the sieve is set aside while the whole final small portion of oversize particles is transferred in small portions to a cooled sapphire percussion mortar contained in an insulated box. Allowing the oversize particles to cool for a few minutes, the reduction in size is effected by percussion and not by grinding (A grinding or rubbing motion will flatten the metal particles making further reduction in size difficult). The powder, considered to be of sufficient fineness, is transferred to a porcelain evaporating dish, a little acetone is added to prevent formation of ice, after which dish and contents are warmed to room temperature by an infra-red heating lamp before sieving the powder; this procedure is repeated until the complete sample has passed a sieve of 200 meshes to the linear inch. Finally the powder –200 mesh is very thoroughly mixed and analysed as a single sample.

One of the problems associated with the use of this technique is that moisture tends to condense onto the mortar during crushing; possibly the use of a glove box may help to eliminate condensation. The powder on completion of crushing is mixed and set aside in a container.

Precaution: Protective gloves should be worn at all times when handling cylinders of liquid air and during the crushing operations as contact of liquid air with the skin will cause very severe burns.

Carbonaceous chondrites

This group of meteorites varies considerably in hardness but is generally able to be crushed by slight pressure in an agate mortar. Care should be taken to ensure that any hard chondrules present in the sample are not lost during crushing. The whole sample is reduced to –100 mesh to the linear inch since no magnetic separation is made.

Silicate phase of stony irons

The silicate inclusions are roughly ovaloid in shape and are best extracted from a cut surface. Usually the solid inclusions are fairly loosely held in the iron matrix and therefore removed by pressure alone. However, if the solid inclusions are difficult to remove, a slight tap with a hardened steel punch will crack the inclusions and so facilitate removal of the silicate material for crushing.

Metal phase of stony irons and iron meteorites

Some metal meteorites, e.g., ataxites, hexahedrites, and the finest octahedrites have a fairly uniform composition. Others may contain inclusions of troilite, cohenite, and schreibersite, by which great care must be exercised if the sample is to be truly representative; if the solid inclusions are large, it is best to analyse these individually in conjunction with an areal analysis of a surface which is as large as possible. Problems are also posed by coarse octahedrites which contain coarse taenite lamellae and stony irons which have relatively large silicate inclusions.

In general, a test sample actually representing the average composition of the material is procured by sawing off, or for drillings by the use of a tungsten carbide-tipped drill. Samples may also be taken by a milling cutter which will give shavings. Initially the surface of the specimen is cleaned free of all extraneous material and given a final wipe with alcohol; if possible, the surface should be polished and etched to assist in recognizing the presence of inclusions or coarse lamellae.

In a study of the composition of eighty eight metal meteorites (Lovering et al., 1957), specimens were sampled by taking random drilling using 1/8-inch high-speed tool-steel drills.

Samples may also be taken from the metal phase of stony irons by carefully placing the point of drilling between the silicate inclusions; however, if silicate inclusions have been touched by the drills, a magnetic separation should remove the metal from the silicate contamination.

Sampling of an individual phase, e.g., schreibersite, may be made using a carbide-tipped dental drill in conjunction with a binocular microscope; although this technique is somewhat tedious, the method does allow a close examination of the inclusion during the drilling operation.

The caution required in choosing the size of sample when sampling coarse octahedrites and siderites, has been emphasized (Moss et al., 1961); it is suggested that a minimum sample size should be calculated depending upon the distance between the taenite lamellae of the sample. The formula given was $2\ t^3$ g, where t is the distance in mm between lamellae. In the general scheme of analysis, a minimum sample weight of 2 g was used for the determination of the major elements. In the special case where a siderite was to be treated with chlorine to remove the metal, the sample was left as a complete piece since the metal phase was continuous.

CORRECTING ANALYTICAL RESULTS TO THE "DRY BASIS"

To secure uniform hygroscopic conditions as a basis for comparison of analytical results, air-dry powder should be used for analysis. With most materials containing hygroscopic water, a special determination of moisture should be carried along at the same time on a separate portion of the air-dry powder by drying at 105–110°C for 1 h, and all other results based on a sample thus dried. To prevent differences of hygroscopicity all the portions necessary for the analysis should be weighed one after the other on the same day.

CHAPTER 8

Silicon

CONSIDERATION OF METHODS

Silicon is, next to oxygen, the most abundant element of the earth's crust; it comprises a good 27.0 per cent of the lithosphere. In nature it does not occur uncombined, but chiefly as silicon dioxide SiO_2 (in quartz, tridymite, opal, and their varieties) or as silicates. The classification of silicates has been dealt with at length by Deer et al. (1963) and the student is referred to this comprehensive study when a detailed account of silicate classification is required.

In the General procedure used by Washington the silicon, as silica, is separated by two evaporations with hydrochloric acid and dehydration, more or less of the small amount of the residual silica inevitably remaining in solution after the double evaporation being caught in the ammonia precipitate in proportion to the size of the precipitate and the amount of silica present, being recovered by appropiate treatments. Where according to Scheme I (p. 85; the one that comes nearest to being in universal use), besides silicon other constituents have to be determined, it would be senseless to determine the residual silica in its combined filtrates of the double (or sometimes recommended triple) evaporations with hydrochloric acid, because the quantity of silica found in these filtrates and washings does not give any insight in the amount of silica which may be carried down by the ammonia precipitate and has to be deducted from the total weight of the Mixed oxides.

If two evaporations with hydrochloric acid have been properly made, a third evaporation will give no more silicon for solubility equilibrium has been reached; the amount of silica remaining in solution (usually not more than 1–2 mg) is almost entirely carried down by the ammonia precipitate because as a rule in silicate analysis a fair-sized precipitate is obtained. Nevertheless, in case of doubt or if very accurate results are desired, the full amount of residual silica that may be still in solution can be estimated satisfactorily by diluting the combined filtrates and washings of the ammonia precipitate to a larger definite volume and a subsequent photometric determination of the silica content by means of the molybdenum blue reaction using an aliquot of the solution; in that case the percentage of silica found in solution should be added to that of the total silica and because of the extraction of the aliquot, a numeral correction must be made where the remainder of the solution is to be used for determination of other constituents. Investigations made (Jeffery and Wilson, 1960a) have shown that the total silica is usually increased by 0.2–0.4% by weight where the residual silica has been measured spectrophotometrically on the filtrate from the separation of the main bulk of silica.

In the General procedure most of the silicon is separated by double dehydrations in acid solution with intervening filtration, and all (or nearly all) of the remainder is caught in the ammonia precipitate, all based on final evaporating with hydrofluoric and sulphuric acids. If these treatments are carelessly done, or if the ammonia precipitate is small, e.g., when iron has preliminarily been eliminated by ether extraction (p. 97) or by the ion-exchange method (p. 109), appreciable amounts of silica may remain in solution and contaminate the oxalate and phosphate precipitates (see pp. 141 and 152). The silica caught in the ammonia precipitate will be counted as alumina if not recovered (see p. 128). Besides this, dehydrated silica is appreciably soluble in acid solution; the amount retained in solution is proportionate to that of the hydrochloric acid used and to the period and temperature of digestion. Its solubility is slightly increased by alkali salts. With hydrochloric acid, dehydration of the silicic acid may be done by evaporating on the steam bath until the residue is dry, after which the residue should be placed in an oven and heated one hour at 110°C. An investigation regarding the optimum conditions for the gravimetric determination of silica

(Anderson, 1962) has shown that dehydration of the silicic acid by heating in an oven at 125°C did not reduce the residual silica. Apart from this investigation, in general less silica (in the absence of magnesium) will pass into the filtrate by exposure to temperatures exceeding 110°C, or after prolonged drying at steam bath temperature, usually at the expense of greater contamination of the filtered silica.

In silicate analysis the contaminants of the filtered silica will vary in composition and in quantity according to the composition of the rock or the mineral. Especially alkali salts, for the very most part derived from fusions with alkali carbonates, are most undesirable contaminants of silica because they volatilize in part during the ignition of the nonvolatile residue, and more or less of the remainder changes in weight during the hydrofluoric-sulphuric acid treatment, thus causing incorrect results for silicon.

Because of the solubility of silica in acid solution on the one hand and the necessity to remove the contaminants on the other hand, the acidity of the solution in which the dehydrated silica is digested or taken up, should be no greater than is necessary for the solution and the removal of accompanying salts and the prevention of hydrolysis (especially where titanium and zirconium are present), whereas the period of digestion for redissolving the salts should be as short as possible. Because of this, the acidity of the solution of the melt in which the silicic acid is digested should not exceed 5–10 % v/v, the period of digestion at steam bath temperature should preferably be not over 10 min and paper and first residue should be washed with sufficient HCl (5 + 95) using hot water for the last washings. The second smaller (usually more coloured) residue obtained after evaporation of the combined filtrate and washings in the same basin should be baked for 1 hour at 110°C, 5 ml of hydrochloric acid and 50 ml of hot water added, the solution allowed to stand on the steam bath for not more than 5–10 min, the solution filtered immediately through a new and smaller close-texture ashless paper, paper and second residue washed with cool HCl (1 + 99) to prevent hydrolysis, and finally carefully with hot water to remove undesirable accompanying salts. Ignition of the silica at approximately 1,200°C for 10–15 min is advisable to drive out all water to obtain constant weight.

As already mentioned, the silica obtained in both residues does not represent all the silicon in the sample for the precipitation of silicon, as silica, is never complete; more or less of the small amount of the residual silicon is to be found and determined in the ammonia precipitate, whereas in case of doubt or in analysis of the highest accuracy, the full amount (or the remainder) of the residual silica that may be still in solution is to be estimated in the combined filtrates and washings of the ammonia precipitate by visual colorimetric or spectrophotometric methods (see p. 78).

The small amount of residual silicon, as silica, caught in the ammonia precipitate, is best recovered in the ignited precipitate by fusing the mixed oxides with pyrosulphate, digestion of the melt in dilute sulphuric acid, evaporating until fumes come off copiously, digesting the cool paste in warm water, collecting the silica on a small close-texture ashless paper, washing paper and residue carefully with hot water.

Due to the amount of time required for evaporations and bakings in the gravimetric method, many attempts have been made to develop a more rapid method for determination of silicon. The spectrophotometric method using the reduced blue silicomolybdate complex has fulfilled this need with certain reservations. In the past, a great deal of controversy has existed over the optimum conditions for the best reproducibility of results but the method given by Shapiro and Brannock (1962) is the one that has now been adopted in the main for the attainment of rapid results. The most appropriate application is when an analysis of a suite of rocks of similar composition is required. Silicon is first determined on two or three typical rocks in the suite by the gravimetric method, using the spectrophotometric procedure for determination of the residual silicon. These samples are subsequently used as standards for the remaining samples. In this manner, interferences have been compensated for by virtue of the similarity of composition.

One of the reservations made for the completely spectrophotometric determination of silicon is that a high degree of experience in analytical technique is required in order to obtain reproducible results; it has been suggested by Churnside (1960) that a good experience of gravimetric work is essential before attempting to apply rapid spectrophotometric methods.

The rapid gravimetric method recommended for the indirect determination of silicon in very highly siliceous material, although exceedingly simple, is affected by some errors; the method should not be applied when accuracy is essential unless it is known that the errors are too slight to be significant or unless modifications are introduced to guard against them. After a preliminary ignition of the sample at about 1,000°C to constant weight (by which sulphides, if present, are converted to oxides), the silicon content is found by treating the ignited sample with hydrofluoric and sulphuric acids, evaporating to volatilize the silicon as the tetrafluoride and to obtain the bases as sulphates, ignition of the residue at 1,000°C to constant weight (by which sulphates such as those of aluminium or iron, if present, are converted to oxides), and subtracting the weight of the contaminants that are left. The loss in weight represents the silica provide that the sulphates were converted completely to oxides, and no silicates happen to be present in the original sample their silicon content being volatilized with the free silica. If desired, nature and amounts of the nonvolatile compunds left after the second ignition can be determined by methods of separation and determination mentioned at various places throughout the book.

GRAVIMETRIC DETERMINATION OF SILICON IN SILICATE ROCKS AND MINERALS IN ABSENCE OF APPRECIABLE CHLORINE, FLUORINE, OR SULPHUR

Consideration of methods (pp. 71–75) should be carefully read before the determination of silicon is undertaken.

According to the General procedure usually followed (Scheme I, p. 85), fuse 1 g of the air-dry sample (–150 mesh to the linear inch) with 4–6 g of pure anhydrous sodium carbonate as described in Chapter 6 (pp. 30–31).

Transfer the fusion cake from the crucible into a 300-ml platinum basin, add some water and if a green colour imparted to the water indicates the presence of manganese (a yellow one chromium, or a bluish-green one the presence of both), for the purpose of reducing manganese, also add some drops of ethyl alcohol. Cover the basin with a cover of platinum or Pyrex glass, and add gradually under the cover 50 ml of HCl (1 + 1). Place the basin on the steam bath and when disintegration is complete, remove the cover and wash any particles or solution adhering to the underside of the cover into the basin with a little water. Evaporate the dissolved carbonate melt and washings in the same basin and allow to stand on the bath for 2 h after approximate dryness, or longer, until the residue is free from fumes of hydrochloric acid. Cover the basin, add cautiously to the dry and cool residue 10 ml of hydrochloric acid, and after 2 min 100 ml of hot water. Remove the cover, wash the inside of the basin with a jet of hot water, heat on the bath for 5–10 min, and stir occasionally until the salts are in solution. Allow to settle somewhat, filter through a wet properly set No.40 Whatman or similar ashless filter paper pouring the solution down a glass rod leaving nearly all the silica at the bottom of the basin. Transfer the main lot of the silica to the paper by means of a jet of hot HCl (5 + 95), wash paper and precipitate 10–12 times carefully and dropwise with hot HCl (5 + 95), then 5 times with hot water allowing to drain well between additions of the washing solutions, and finally suck dry at the pump. (It is very important that the silica and paper be washed free from acid and contaminating salts, see p. 73.) Reserve paper and residue (*1*).

Next evaporate the combined filtrate and washings (*2*) to dryness in the same basin, transfer basin and contents to an oven, and bake for 1 hr at 110°C. Cover, drench the residue with 5 ml of hydrochloric acid, add 50 ml of hot water, stir, allow to stand on the bath for 5–10 min, filter immediately through a new and smaller wet properly set No.42 Whatman or similar ashless filter paper, transfer the precipitate to the paper by means of a jet of cool HCl (1 + 99), wash paper and precipitate 10–12 times carefully and dropwise with cool HCl (1 + 99), and finally 5 times with hot water as before; the film of silica adhering to the basin should be detached by means of a small piece of ashless filter paper wrapped round a rubber-tipped glass rod and added to the precipitate. (Cool dilute acid prevents hydrolysis and removes impurities from the small amount of silica which is never pure and more coloured than the first one. After washing paper and precipitate with hot water, complete removal of chloride ions may be tested by collecting a few drops from the last washing in a test tube containing 2 ml of an 0.1 *N* solution of silver nitrate and 1 drop of 4 *N* nitric acid; the white curdy precipitate of AgCl is entirely soluble in ammonia.) Reserve paper and residue (*3*), and also the combined filtrate and washings (*4*).

Next transfer the two papers with their contents (*1* and *3*) to a platinum crucible weighed with well-fitting cover, moisten with some drops of H_2SO_4 (1 + 4), carefully dry the papers and contents and char the papers without inflaming.

Burn the carbon under good oxidizing conditions at as low a temperature as possible, cover the crucible, and using a Meker burner finally heat at approximately 1,200°C for 15–20 min. (The heating should be very gently at first to avoid loss of finely crystallized silica by drafts, and the crucible must be tightly covered during the final heating.) Allow to cool in a small desiccator over a good desiccant, weigh without delay while still covered, and repeat the ignition and weighing until constant weight is obtained.

Then raise cover slightly, carefully moisten the contents with water by means of a small pipette inserted against the side of the crucible, add some drops of H_2SO_4 (1 + 1), and wash any particles or solution adhering to the underside of the cover into the crucible with a little water (Sulphuric acid should be added to convert all the contaminating bases to sulphates and to prevent loss of titanium or zirconium as fluorides).

Next add approximately 10 ml of hydrofluoric acid (free from nonvolatile impurities) delivered directly from the polyethylene or ceresin bottle, evaporate slowly upon the steam bath under a good hood to volatilize the silicon as silicon tetrafluoride, cover the crucible, carefully expel the sulphuric acid, and ignite at approximately 1,000°C for 1–2 min. (This temperature is high enough to change the nonvolatile matter to the state in which it occured in the impure silica.)

Allow to cool in the small desiccator, weigh without delay while still covered, and repeat the ignition and weighing until constant weight is obtained; the difference between the two weights represents the weight of the main lot of silica (*6*). To this must be added:

(*a*) the weight of the amount of Residual silica (*14*) recovered from the impure Mixed oxides (*13*);

(*b*) the weight of the full amount (or of the remainder) of the Residual silica (estimated by colorimetric or spectrophotometric methods) still left in solution in filtrate (*12*) when the ammonia precipitates (*9* and *11*) are small, or when the iron or aluminium have preliminary been eliminated, in which cases a numeral correction must be made where the remainder or filtrate (*12*) is to be used for determination of other constituents (see p. 81). In all cases where very accurate results are desired, corrections for impurities by attack of the vessels or derived from the reagents are also to be determined.

Reserve the platinum crucible containing the nonvolatile residue (*5*) from the main lot of silica (*6*) for subsequent determinations.

SPECTROPHOTOMETRIC DETERMINATION OF SILICON

Determination of residual silicon in filtrate (12) (see Scheme I)

Principle of method

Formation of yellow silicomolybdate in weak acid solution followed by its reduction to molybdenum blue at a degree of acidity sufficiently high to suppress interference. Photometric measurement is made at approximately 650 mμ.

Concentration range

The recommended concentration range is from 0.05 to 0.2 mg of silicon in 100 ml of solution, using a cell depth of 1 cm.

Stability of colour

After the addition of the ammonium molybdate solution the full colour develops in 10 min. Addition of the oxalic acid should be made immediately after this period as its addition stops further formation of the complex. A uniform time for colour development should be used for both calibration solutions and samples. The colour is stable for about two hours.

Interfering elements

The elements that may be present in filtrate (*12*) do not interfere or their concentrations are so low that their interference can be tolerated in the procedure given.

Apparatus

Borosilicate glass volumetric ware should be used and samples transferred to absorption cells just prior to reading. Standard volumetric flasks, burettes, and pipettes, should be of precision grade (see p. 5). All glassware should be rinsed with HCl (1 + 1) before use. Solutions should be transferred at once to a plastic or hard rubber bottle, or prepared fresh as needed. Funnels of plastic, or hard rubber, about 6 cm in diameter may be used, fitted with a plastic tubing about 8 mm in diameter cemented on to lengthen the stem for dipping into the solutions previously added to the flasks.

Special reagents

(*a*) Standard silicon solution (1 ml ≡ 0.05 mg Si). – Fuse

0.1070 g of pure anhydrous silica (as pure quartz crystals, or "spec-pure" silica) with 1.0 g of anhydrous sodium carbonate in a platinum crucible. Cool the melt, dissolve completely in water, transfer the solution to a 100-ml volumetric flask, dilute with water to the mark and mix. Next transfer a 5.0-ml aliquot of this solution to a 250-ml volumetric flask containing 200 ml of water and 20 ml of HCl (1 + 1), dilute with water to the mark, and mix. Transfer at once to a plastic or hard rubber bottle.

(*b*) Ammonium molybdate solution (100 g $(NH_4)_6Mo_7O_{24}.4H_2O$ per liter). – Dissolve 10 g of ammonium heptamolybdate in 60 ml of water and 6.0 ml of ammonium hydroxide, and dilute to 100 ml with water.

(*c*) Oxalic acid solution (100 g $H_2C_2O_4.2H_2O$ per liter). – Dissolve 10 g of crystallized oxalic acid in water and dilute to 100 ml.

(*d*) Reducing solution. – Dissolve with shaking 0.15 g of 1,2,4-aminonaphtol-sulphonic acid, 0.7 g of sodium sulphite and 9 g of sodium metabisulphite in 100 ml of water. Prepare this solution fresh monthly and store in refrigerator.

Preparation of calibration curve

Transfer 1, 2, 3, 4 and 5 ml of silicon solution (reagent *a*) to 100-ml volumetric flasks. To each flask and to an additional flask for the blank, add 5 ml of water and 1.0 ml of ammonium molybdate solution (reagent *b*), mix, and allow to stand for 10 min. Then add 5.0 ml of oxalic acid solution (reagent *c*), mix, add 2.0 ml of reducing solution (reagent *d*), mix, dilute with water to the mark, mix, and allow to stand for 30 min.

Next transfer a suitable portion of the calibration solution to an absorption cell having a 1-cm light path. Using a spectrophotometer measure the absorbancy at approximately 650 mμ, compensate or correct for the blank, and plot the values obtained against mg of silicon per 100 ml of solution.

Procedure

Transfer filtrate (*12*) to a 100-ml beaker, acidify with HCl (1 + 1), add some drops of bromine water, boil the solution to destroy the indicator and to remove the excess of bromine, cool to 20°C, transfer the solution to a 250-ml volumetric flask, add 20 ml of HCl (1 + 1), dilute with water to the mark, and mix.

Next transfer a 5.0-ml aliquot of this solution to a 100-ml volumetric flask. To the flask, and to an additional flask for the blank, add 5 ml of water and 1.0 ml of ammonium molybdate solution (reagent *b*), mix, and continue in accordance with "Preparation of calibration curve". Compensate or correct for the blank.

Using the value obtained, read from the calibration curve the number of mg of silicon present in 100 ml of solution, calculate the amount of the silicon present in filtrate (*12*), and add the weight of the full amount, calculated as Residual silica (SiO_2 = 2.14 × Si), to that of the main lot of SiO_2 (*5*).

Determination of silicon in a suite of rocks of similar composition

Principle of method

Silicon is first determined on two or three typical rocks in the suite by the gravimetric method, using a spectrophotometric procedure for the determination of the Residual silicon. These rocks resembling as closely as possible in chemical and physical nature the remaining rocks, are used as standards; errors arising from different factors affecting both samples and standards alike, will be compensated by virtue of the similarity of composition. The photometric method used, is based on the measurement of the absorbancy of the blue colour produced by reduction of the yellow silicomolybdate formed in weak acid solution at a degree of acidity sufficiently high to suppress interference. Photometric measurement is made at approximately 640 mμ.

Concentration range

The recommended concentration range is from 0.1 to 0.4 mg of silica in 100 ml of solution, using a cell depth of 1 cm.

Stability of colour

After the addition of reducing solution and diluting with water to the mark, the full colour develops in 30 min. A uniform time for colour development should be used for both standard solutions and samples. The colour is stable for about two hours.

Interfering elements

Similarity of composition of standards and samples will compensate errors arising from different factors. The presence of iron decreases the absorbance but this effect can be substantially diminished by allowing the reduced solution to stand for 1 h. If this procedure is followed, a quantity of up to 3 mg of ferric iron in 100 ml of solution does not interfere (Tuma, 1962).

Apparatus

Although fusions with sodium or potassium hydroxide should preferably be made in crucibles of gold or silver, and at a lower temperature than is tolerated with platinum (see p. 10), the decomposition of silicates can be carried out by molten NaOH or KOH at about 400°C in a nickel crucible. For general specifications, etc., consult p. 9.

Special reagents

(*a*) Calibration solutions. – The solutions are obtained by treating two or more standard rocks of known composition (or standard analyzed samples distributed by, e.g., the National Bureau of Standards) at the same time and under identical conditions, presuming that the values obtained for the unknown rocks are equally accurate.

(*b*) Sodium hydroxide solution (30 g per 100 ml). – Dissolve 30 g of sodium hydroxide in a plastic beaker in water and dilute to 100 ml. Transfer at once to a plastic or hard rubber bottle.

(*c*) Ammonium molybdate solution (7.5 g $(NH_4)_6Mo_7O_{24}.4H_2O$ per 100 ml). – Dissolve 7.5 g of ammonium heptamolybdate in 75 ml of water and 25 ml of H_2SO_4 (1 + 4).

(*d*) Tartaric acid solution (10 g $C_4H_6O_6$ per 100 ml). – Dissolve 10 g of tartaric acid in water and dilute to 100 ml.

(*e*) Reducing solution. – For preparation, consult p. 80.

Procedure

Because of the fact that sodium hydroxide gives off water by fusion, accompanied by frothing and spitting, transfer 5 ml of the sodium hydroxide solution (reagent *b*) by means of a small graduated plastic cylinder to each of four nickel crucibles of 75-ml capacity with snugly fitting covers (viz. two crucibles for the standards, one for the sample under test, and one for the blank), evaporate the solution to dryness, bring the residues into quiescent fusion avoiding direct contact with the flame, and allow the melts to solidify and to cool.

Weigh 50–100 mg of the air-dry sample (–60 mesh to the linear centimeter) to the nearest 0.1 mg on a suitable cover glass, pour as much as possible of the powder upon the cool melt, reweigh the cover glass and note the weight of the portion transferred; repeat this method of weighing for the two standard rocks of known composition (or for standard analyzed NBS-samples).

Cover the crucibles, heat to dull red heat for ca. 5 min avoiding direct contact with the flame, next carefully remove covers, swirl the melts gently just as they are cooling so that thin layers are formed on the walls of the crucibles, and allow the crucibles to cool to room temperature.

Fill the well-cooled crucibles with ca. 50 ml of water, replace the covers and allow to stand until the melts have dissolved, which may be overnight. (Disintegration of the melts is aided by stirring occasionally with plastic stirring rods.)

When disintegration is complete, transfer the contents of each crucible separately to a 1-l volumetric flask containing ca. 400 ml of water and 20 ml of HCl (1 + 1), rinse the crucibles well with water to ensure complete transfer of the contents, dilute with water to the mark, and mix.

Transfer an aliquot of the four solutions (e.g., 10 ml) separately to a 100-ml volumetric flask, add ca. 50 ml of water and 2.0 ml of ammonium molybdate solution (reagent *c*) to each flask, mix, and allow to stand for 10 min. Then add 4.0 ml of tartaric acid solution

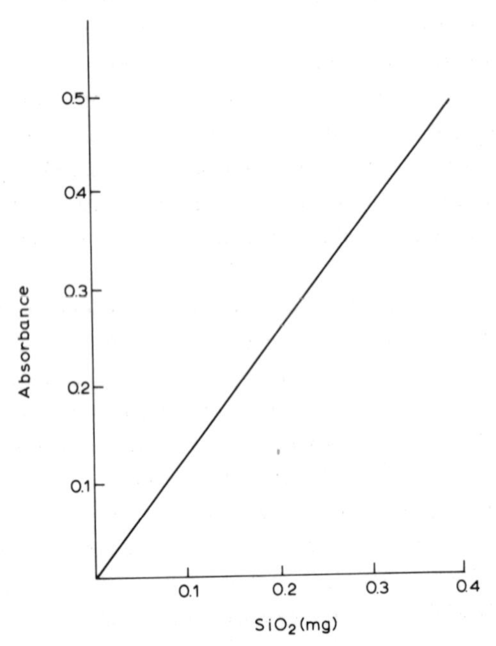

Fig.7. Calibration curve for silica using a 1-cm cell, 100 ml volume and a wavelength of 640 mμ.

SCHEME I

General procedure for oxidized materials in absence of appreciable chlorine, fluorine or sulphur

Determination of SiO_2, Total iron as Fe_2O_3, Al_2O_3, CaO, and MgO

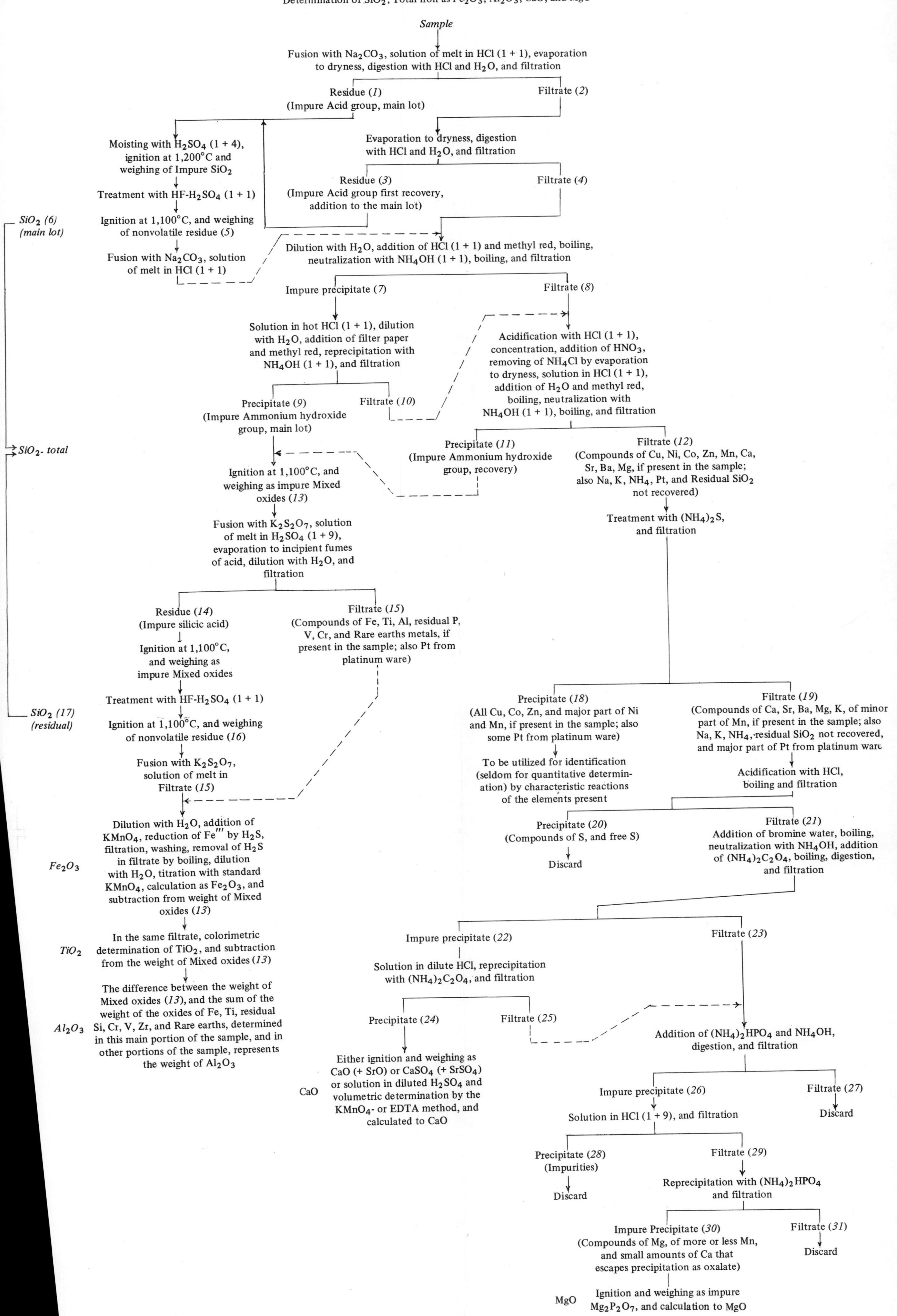

(reagent *d*), mix, add 1.0 ml of reducing solution (reagent *e*), mix, dilute with water to the mark, mix, and allow to stand for 30 min.

Next transfer a suitable portion of the coloured solutions separately to an absorption cell having a 1-cm light path. Using a spectrophotometer measure the absorbancies at approximately 640 mμ, compensate or correct for the blank, and plot the value given by the sample under test, against the values given by the calibration solutions (reagent *a*).

Fig.7 shows an approximation of the values obtained within the concentration range of 0.1–0.4 mg of SiO_2 in 100 ml of solution, using a cell-depth of 1 cm.

VOLUMETRIC DETERMINATION OF SILICON

McLaughlin and Biskupski (1965) have described a modified procedure for the precipitation of silicon as potassium fluorosilicate from a sodium peroxide fusion, Interference in this method from aluminium and titanium is suppressed by the addition of calcium chloride. The solubility of the precipitate is reduced by use of a low pH and an excess of potassium chloride (cf. Silicon, p. 73). The precipitate is then titrated with a standard solution, phenolphthalein being used as an indicator.

RAPID GRAVIMETRIC DETERMINATION OF SILICON IN HIGHLY SILICEOUS MATERIALS

Decomposition by acids (pp. 35–38) should be carefully read before the determination of silicon is undertaken.

In highly siliceous materials silicon, as silica, may be simply and rapidly determined by the loss in weight that occurs when these materials are treated with hydrofluoric and sulphuric acids; the loss of weight may be taken to be the weight of silica with close approach to accuracy, provided: (*a*) the volatilization of silicon, as silicon tetrafluoride, has been preceded by ignition of the air-dry sample at about 1,000°C to constant weight, at which temperature volatile substances will be removed and sulphides and certain sulphates, if present, will be converted to oxides which do not interfere; (*b*) the weight of the ignited impurities is small and does not change as a result of the treatments between weighings.

Ignite up to 1 g of the air-dry sample (–200 mesh to the linear inch) in an open 100-ml well cleaned, well ignited, and cool platinum dish equipped with a platinum cover for 5 min at 1,000°C in an electrically heated muffle controlling the temperature. Cover the dish, cool in a desiccator containing colourless sulphuric acid, and repeat the operation until constant weight is obtained.

Next moisten the ignited sample with water, add about 10 ml of hydrofluoric acid (free from nonvolatile impurities) by means of a small graduated plastic cylinder, evaporate slowly with occasional stirring with a platinum rod upon the steam bath under a good hood to dryness, and repeat the operation with fresh hydrofluoric acid until all gritty particles have disappeared. (Silicate minerals are easier decomposed by HF than crystallized silica.)

After the final evaporation, add some drops of sulphuric acid, by increasing the temperature carefully evaporate the residue to dryness, cover the dish, subject the dry residue to the same degree of heat (viz. 1,000°C) as in the first ignition, cool and weigh as before; the last weighing gives the weight of silica volatilized, and of the impurities as oxides, if the sulphates were completely decomposed. (The composition of the residue left after the final ignition can be determined by the usual methods.)

CHAPTER 9

Total Iron

CONSIDERATION OF METHODS

Iron is, next to aluminium, the most abundant metal of the earth's crust; being a constituent of numerous rocks and minerals it comprises about 5.0 per cent of the lithosphere. Native iron occurs as meteoritic iron (always alloyed with nickel, and accompanied by small amounts of other metals and occluded gasses) in most meteorites.

Disregarding sulphides, iron may occur in silicates in both the ferrous and ferric state. If present in both states, it is customary to determine the Total iron as ferric iron in filtrate (*15*) of the main portion as outlined in Scheme I, and the Ferrous iron in another portion of the sample (see p. 161). When sulphides are absent and the amount of the ferrous iron has been determined in addition to the total iron in terms of Fe_2O_3, from this last figure the value of the ferric oxide equivalent of the ferrous iron has to be deducted to obtain the value for the ferric oxide equivalent of the ferric iron. When decomposable sulphides are present, whether ferro compounds or not, it is difficult and sometimes impossible to obtain absolutely reliable results because two errors may arise, viz. one from an iron content of the sulphides, and another from the hydrogen sulphide liberated in dissolving the sample for the ferrous or ferric oxide determination by which procedures more or less ferric iron will be reduced and titrated in the ferrous state, thus causing high results for the ferrous iron. Also very small amounts of vanadium (as may be present in silicate rocks) in dilute sulphuric acid solution reduced by e.g., hydrogen sulphide to the quadrivalent state, will affect the volumetric or colorimetric reduction–oxidation methods and cause high results for the reduced iron compounds.

In silicate analysis the most widely adopted methods for the determination of iron are volumetric, most gravimetric methods (by weighing as oxide) being too subject to interference; e.g., a direct determination of iron by precipitation with reagents such as ammonium hydroxide or cupferron is rarely possible because these procedures leave the iron contaminated with other precipitable elements.

Interchange between the ferrous and ferric states of oxidation (processes in which transfers of electrons take place and corresponding changes of valence occur among atoms or groups of atoms) is the basis of volumetric methods, many reductants (as stannous chloride, sulphur dioxide, hydrogen sulphide, titanous salts, nascent hydrogen) and oxidants (as potassium permanganate, potassium dichromate, ceric sulphate) being in use, the end point being found potentiometrically or by means of an indicator. Any given reduction-oxidation method, or oxidation-reduction method, for the determination of iron requiring different conditions, may be subject to interference by certain elements or compounds, and as a rule requires their preliminary elimination.

In silicate analysis it is customary to obtain the total iron in the reduced state as ferro compound and to measure the re-oxidation to the ferric state (potentiometrically, or by using an internal or external indicator) by means of standard oxidizing solutions (of e.g., potassium permanganate, potassium dichromate, or ceric sulphate), the adoption of oxidation-reduction methods involving direct oxidation and measured reduction by the use of standard reducing solutions (of e.g., titanous or chromous salts) being limited by the fact that the usual standardized reductants, in particular solutions of titanous salts, are extremely unstable, easily oxidized by air, and require storage under an inert gas such as purified hydrogen, carbon dioxide or nitrogen.

For determination of iron in silicates, modified volumetric, electrolytic and colorimetric methods have been devised using special techniques, as e.g., reduction in silver reductor, separation by chromatographic adsorption, separation by ion-exchange, and spectrophotometric methods (usually confined to small amounts) based on the formation of soluble coloured complexes of iron with o-phenanthroline, thiocyanate, salicylate, bipyridine, dipyridyl, etc. Snell and Snell (1959) list 26 reagents for the photometric determination of iron; many more are now available. Recently a method using the coloured EDTA-iron complex has been introduced (Poeder et al., 1962).

REMOVAL OF IRON BY EXTRACTION WITH ETHER

The separation of iron from other elements, based on the ether extraction of ferric chloride preferably from a cold dilute hydrochloric acid solution, is used only if iron is present in such large amount that it interferes in determinations of the accompanying elements. Although in this procedure, converted to certain valences, elements such as aluminium, titanium, manganese, chromium, nickel, cobalt, beryllium, calcium, magnesium, thorium, zirconium, and rare earths, will remain as chlorides in the acid layer. The extraction of ferric chloride (even after repeated treatments) is never quantitative as its separation from elements such as vanadium, copper, platinum, phosphorus, zinc and tin, is incomplete.

Using this procedure the ferric chloride present in cold hydrochloric acid solution having a specific gravity of 1.10, is extracted by ether that should be free from alcohol and from substances which decompose the ether (e.g., free chlorine or nitric acid), as well as from salts insoluble in hydrochloric acid saturated with ether (e.g., alkali chlorides), and other acids (such as sulphuric acid) that will retard or lower the extraction of the ferric iron. Moreover, the separation should be carefully checked in case other chlorides, which may be of interest, are also extracted. (For further details of the separation method, see Swift, 1924.)

DETERMINATION OF TOTAL IRON IN SILICATE ROCKS AND MINERALS

Common operations in silicate analysis (Ammonium hydroxide group, p. 51) should be carefully read before the determination of Total iron is undertaken.

According to the general procedure (p. 85), fuse the nonvolatile residue from the main lot of silica retained in the platinum crucible (*5*) with about 0.3 g of sodium carbonate, cool, dissolve the melt in a slight excess of HCl (1 + 1), and add the solution to filtrate (*4*) from the determination of silica. Dilute this filtrate to ca. 180 ml with water, add 20 ml of HCl (1 + 1) and a few drops of methyl red indicator (0.2% alcoholic solution), heat just to boiling, while stirring with a rubber-tipped glass rod add NH_4OH (1 + 1) dropwise until the red colour of the solution just changes to a light yellow. Boil the solution for not longer than 2 min, add one or more drops of NH_4OH (1 + 1) if the red colour of the indicator reappears, filter at once and rapidly through a wet properly set No.40 Whatman or similar ashless filter paper, wash the impure precipitate (*7*) 6–8 times with a hot neutral ammonium chloride solution (2%), and reserve filtrate (*8*).

Next place funnel with contents above the beaker in which the precipitation was made, unfold the paper, place it against the inside of the funnel, return the precipitate to the beaker by washing it off the paper with a strong jet of boiling water and dissolve the precipitate by adding 25 ml of hot HCl (1 + 1).

Dilute the solution to ca. 200 ml, add the paper, stir until pulped, repeat the precipitation and washing as before, and reserve filtrate (*10*). Wrap the washed precipitate (*9*) in its paper, reserve in a platinum crucible weighed with well-fitting cover, and combine first and second filtrates and washings (*8* and *10*) in a suitable beaker.

In order to avoid difficulties offered by large amounts of ammonium chloride present in solution, acidify the combined filtrates and washings (*8* and *10*) with HCl (1 + 1), evaporate to small volume, cool, cover the beaker, add at least 2.5 ml of nitric acid (an excess does not harm) for every gram of ammonium chloride present, warm the mixture on the bath until evolution of chlorine has ceased, wash and remove the cover, and finally evaporate to dryness. Next add ca. 10 ml of HCl (1 + 1), evaporate to dryness to convert all nitrates into chlorides, cool, add 5 ml of HCl (1 + 1), 40 ml of water and one drop of methyl red indicator, heat just to boiling, repeat the precipitation and washing as before, collect precipitate (*11*) on a smaller ashless paper and reserve filtrate (*12*).

Transfer the washed small paper and precipitate to the platinum crucible containing the reserved main precipitate (*9*), and dry the papers with contents on the bath. Using a Meker burner, char the papers without inflaming, burn the carbon at a low temperature and under good oxidizing conditions, next cover the crucible and heat for 5 min at approximately 1,100°C. Allow to cool in a small desiccator over colourless sulphuric acid, weigh without delay while still covered, and repeat the ignition and weighing of the impure Mixed oxides (*13*) until constant weight is obtained.

Add a known amount (about 5–7 g) of potassium pyrosulphate to the crucible, fuse the impure Mixed oxides in the covered crucible at a low temperature until the whole has been dissolved, and pour the melted mass as much as possible into a large well-ignited and cool platinum dish. Next measure 100 ml of H_2SO_4 (1 + 9), bring the small amount of melt adhering to the crucible into solution by digestion on the bath with a part of the measured dilute acid, transfer the solution and washings to the dish, add the remainder of the 100 ml dilute acid, digest until the solid is dissolved, increase the heat and evaporate to incipient fumes of sulphuric acid.

Cool, to the pasty mass add 100 ml of warm water, heat on the bath until the salts are in solution and silica has coagulated, and immediately filter through a small wet properly set No.42 Whatman or similar ashless paper into a 750-ml flask marked at the 100 ml and 400 ml level. Wash the small residue carefully with hot water, transfer the washed paper and residue (*14*) to a 10-ml platinum crucible weighed with well-fitting cover, and reserve filtrate (*15*).

Dry the small paper and residue on the bath, char the paper

without inflaming, burn the carbon at a low temperature and under good oxidizing conditions, next cover the crucible and heat for 2 min at approximately 1,100°C. Cool in desiccator, weigh while still covered, and repeat the ignition until constant weight is obtained. Treat the weighed small residue with 1 ml of hydrofluoric acid and 2 drops of H_2SO_4 (1 + 1), carefully evaporate to dryness, ignite, cool, and weigh again. The loss in weight by the HF-H_2SO_4 treatment, representing the weight of Residual silica carried down by the ammonia precipitates (*9* and *11*) must be added to that of the main lot of SiO_2 (*6*) already obtained.

In case of doubt or if very accurate results are desired, the full amount (or the remainder) of the Residual silica that may be still in solution, can be recovered and estimated in the combined filtrates and washings of the ammonia precipitate (*12*); the calculated weight of the amount of Residual silica (*17*) must also be added to that of the main lot of SiO_2 (*6*) to arrive at the total amount of SiO_2 (*18*) present in the sample.

Fuse residue (*16*) in the crucible with a small amount of potassium pyrosulphate, dissolve the cooled melt in filtrate (*15*) and reserve the filtrate for subsequent determinations.

VOLUMETRIC DETERMINATION

Total iron by the sulphide-permanganate method without regard to vanadium (see Scheme I)

Principle of method

The ferrous sulphate obtained after treatment with hydrogen sulphide in dilute sulphuric acid solution, is in cold solution oxidized by a standard potassium solution as quantitative oxidant. In silicate analysis the method is almost always applied because of its selectivity; even very dilute solutions of ferrous iron can be titrated accurately provided the specified technique is used. Members of the hydrogen sulphide group (very little, if any) present in the Mixed oxides are precipitated by the treatment and cause no trouble; notable amounts of platinum metals introduced through actions on the crucibles and dishes may be removed by filtration before expelling the hydrogen sulphide. Titanium and chromium (nearly always present in the Mixed oxides) do not interfere. Vanadium interferes because it is reduced to the quadrivalent state by hydrogen sulphide and is later oxidized to the quinquevalent state in titrations with permanganate.

The actions of hydrogen sulphide and potassium permanganate on dilute sulphuric acid solutions of iron (III) and vanadium (V) are:

$$Fe_2(SO_4)_3 + H_2SO_4 + H_2S \rightarrow 2\,FeSO_4 + 2\,H_2SO_4 + S$$

$$10\,FeSO_4 + 2\,KMnO_4 + 8\,H_2SO_4 \rightarrow 5\,Fe_2(SO_4)_3 + 2\,MnSO_4 + K_2SO_4 + 8\,H_2O$$

$$2\,H_3VO_4 + 2\,H_2SO_4 + H_2S \rightarrow V_2O_2(SO_4)_2 + 6\,H_2O + S$$

$$5\,V_2O_2(SO_4)_2 + 2\,KMnO_4 + H_2SO_4 + 22\,H_2O \rightarrow 10\,H_3VO_4 + 2\,MnSO_4 + K_2SO_4 + 8\,H_2SO_4$$

Special reagents

(*a*) Chromic acid solution, saturated. – Dissolve 20 g of $K_2Cr_2O_7$ in 1 liter of water, add slowly while stirring and cooling an equal volume of H_2SO_4. When cool, store in a 2-l glass-stoppered bottle.

(*b*) Hydrogen sulphide wash solution, freshly prepared. – Saturate H_2SO_4 (1 + 99) with H_2S of high purity.

(*c*) Lead acetate paper.

(*d*) Sulphuric acid (5 + 95), freshly boiled. Boil for 10–15 min and cool to room temperature.

(*e*) Potassium permanganate solution (20 g per l). – Dissolve 2 g of $KMnO_4$ in 100 ml of water.

(*f*) Sodium oxalate (NBS standard sample No.40c, or pure sodium oxalate prepared according to Sörensen). – Dry for 1 h at 105°C.

(*g*) Water, free from reducing substances. – Distill water from an alkaline permanganate solution.

(*h*) Standard potassium permanganate solution (0.1 *N* or less). – Because of the fact that freshly made solutions of permanganate lose strength at first through action between the permanganate and organic matter in the water or containing vessel or impurities in the permanganate themselves, more stable standard solutions can be obtained by preparing solutions of slightly more than the desired strength and to standardize them after boiling and cooling. More dilute solutions (e.g., 0.02 *N* or less) should always be prepared by diluting aged 0.1 *N* solutions with water that has been freed from reducing substances.

Preparation of standard $KMnO_4$ solution (0.1 *N*): transfer 3.2 g of $KMnO_4$ and 1 l of water to a 1.5-l flask, boil gently (protected from dust and reducing vapors) for 1 h, cover, and let stand in the dark overnight. At the same time fill a 1-l glass-stoppered bottle to the neck with saturated chromic acid solution (reagent *a*), plunge a suitable fritted-glass filtering funnel of medium porosity into a 600-ml beaker filled with the same saturated solution, and let stand overnight. Next return the chromic acid solution to the original 2-l bottle, rinse the 1-l bottle and the beaker well with water, cover the bottle with black paper, wash the filtering funnel thoroughly with water and drain under full suction. Insert the stem of the funnel into the neck of the black-covered bottle, filter the $KMnO_4$ solution directly into the bottle without aid of suction, stopper, and keep the solution away from light; the oxidimetric value of this solution is approximately 0.1 *N* with respect to sodium oxalate.

Standardization of the standard $KMnO_4$ solution against sodium oxalate (Fowler and Bright, 1935): accurately weigh and transfer 0.3000 g of the dried sodium oxalate (reagent *f*) to the 600-ml beaker, and 250 ml of freshly boiled and cooled dilute sulphuric acid (reagent *d*), and stir until the oxalate has dissolved. While stirring slowly, introduce 39–40 ml of the ca. 0.1 *N* $KMnO_4$ solution

(0.3000 g of $Na_2C_2O_4$ requires 44.77 ml of 0.1 *N* $KMnO_4$) from a 50 ml burette at a rate of 25–30 ml per min. Let stand until the pink colour disappears, this generally taking about 45 sec. (If too much permanganate has been added, discard, and begin again, adding a few ml less of the $KMnO_4$ solution.) Heat to 55–60°C, and complete the titration by adding the $KMnO_4$ solution until a faint pink colour persists for 30 sec. Add the last 0.5–1 ml drop by drop, with particular care to obtain complete reduction of each drop before the next is introduced. Determine the amount of $KMnO_4$ required to impart a faint pink colour to the solution by adding the $KMnO_4$ solution to 250 ml of the freshly boiled and cooled dilute sulphuric acid (reagent *d*) at 55–60°C, and subtract this correction (usually 0.03–0.05 ml) from the observed titration. Calculate the normality and the titer of the standard solution. One ml of an exactly 0.1 *N* $KMnO_4$ solution is equivalent to 5.585 mg of Fe (bivalent to trivalent state), and to 5.095 mg of V (quadrivalent to quinquevalent state).

Allow the standard solution to stay in the burette for only a short time, and never return unused portions of the solution to the container. Carefully prepared, 0.1 *N* $KMnO_4$ solutions stored in the dark, suffer no appreciable decrease in titer for several months.

The method as described here, can be also followed in preparing other desired concentrations; however, $KMnO_4$ solutions ($<0.1\,N$) decompose more rapidly and should be restandardized frequently.

Procedure

When the amount of iron is too small for satisfactory titrations, and no suitable adjustments can be made in order to produce the desired concentration, the percentage of Total iron in a rock or mineral should be determined by the use of more refined techniques as e.g., by spectrophotometric methods.

The titrimetric determination of the Total iron in the sample should be carried out as follows:

Dilute filtrate (*15*) with water to 400 ml, heat to boiling, add

$KMnO_4$ solution (reagent *e*) drop by drop until a distinct pink colour persists and, performed in a good hood, pass a rapid stream of H_2S of high purity through the solution for 30 min. Remove the flask from the source of heat, continue the stream for 15 min and if sulphides are precipitated, filter the solution through a suitable fritted-glass filtering funnel without aid of suction, into another 750-ml flask marked at the 100 ml level, and wash the precipitate moderately with H_2S wash solution (reagent *b*). To prevent bumping, add five solid glass beads having a diameter of about 5 mm, boil the filtrate and washings vigorously until reduced to a volume of 100 ml to expel the H_2S (completeness of removal of H_2S may be tested with lead acetate paper, reagent *e*), rinse down the neck of the flask with water, boil the solution for 3 min, cool quickly to room temperature (oxidation of ferrous iron during the cooling of the solution is negligible), dilute with 100 ml of non-reducing water (reagent *g*) and titrate the iron immediately with the standard $KMnO_4$ solution (reagent *h*) until one drop produces a slight pink coloration which persists without fading for at least 30 sec. (When vanadium is present, the colour does not persist until the vanadyl salt is completely oxidized; in cool solution, however, it reacts very slowly. Make a blank determination following the same procedure and using the same amounts of all reagents.

$$\text{Total iron \%} = \frac{(A - B)N \times 0.0559}{C} \times 100$$

where:

A = ml of standard $KMnO_4$ solution required for titration of the sample:

B = ml of standard $KMnO_4$ solution required for titration of the blank;

N = normality of the standard $KMnO_4$ solution;

C = grams of sample used in titration.

SPECTROPHOTOMETRIC DETERMINATION

Total iron as Fe_2O_3 by the 2,2′-dipyridyl method
(see Scheme II)

When the Total iron content of the sample is too low for satisfactory titrations (see Procedure, p. 103) a very useful method for the photometric determination of small amounts of iron in silicate material is that based on the formation of a coloured ferro dipyridyl complex (Smith et al., 1958), since fluorides, phosphates, tartrates, oxalates, etc., do not interfere. The method is usually applied to an

SCHEME II

General procedure for dissolution of the sample by acids

(1) transfer 0.5–1.0 g of air-dried sample to platinum basin, add 25 ml water + 20 ml HF + (1 ml H_2SO_4 or other acid)

↓

(2) digest covered for 30 min on water bath

↓

(3) remove cover and evaporate to fumes

↓

(4) cool, repeat evaporation after washing the inside surface of the basin with H_2SO_4 (1 + 9)

↓

(5) evaporate to fumes

↓

(6) cool, add 50 ml water and dissolve salts, warm if necessary

↓

(7) cool, transfer solution to volumetric flask

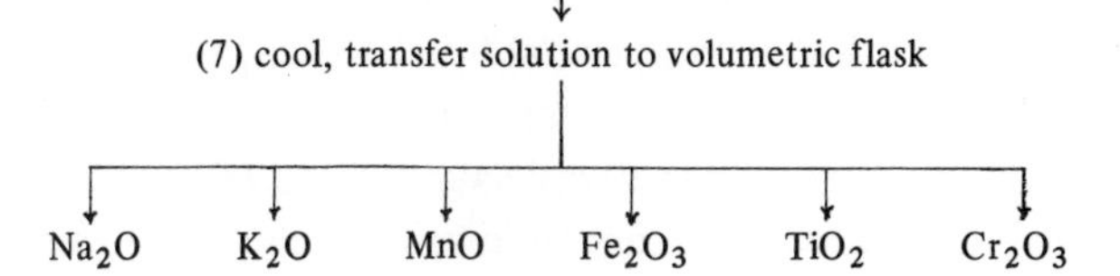

aliquot of the clear solution of the soluble bases obtained after decomposing silicates by the use of hydrofluoric and sulphuric acids as described on p. 35. Any residual particles resistant to the acid attack must be separated, fused with potassium bisulphate, the melt leached out with dilute acid and the soluble bases combined with those in the main solution; using this procedure, most of the silica still present remains undissolved along with the alkaline earths, according to their nature and amounts.

Principle of method

Ferrous iron in a solution having a pH of about 5.5, forms a reddish-purple complex with 2,2′-dipyridyl. Photometric measurement is made at approximately 522 mμ.

Concentration range

The recommended concentration range is from 0.05 to 0.5 mg of Fe_2O_3 in 100 ml of solution, using a cell depth of 1 cm.

Stability of colour

After the addition of the 2,2′-dipyridyl solution the full colour develops within five minutes. Addition of the sodium acetate should be made after formation of the complex. A uniform time for colour development should be used for both calibration solutions and samples. The colour is stable for 24 hours.

Interfering elements

If the separations recommended in the General procedure (Scheme I) are made properly, none of the elements present will interfere. Elements ordinarily present in silicate rocks or minerals, obtained as soluble sulphates after decomposition of the sample by the use of hydrofluoric and sulphuric acids (Scheme II), interfere little if any.

Special reagents

(*a*) Standard iron solution (1 ml ≡ 0.01 mg Fe_2O_3). – Dissolve 0.1398 g of "specpure" iron sponge in the minimum quantity of hydrochloric acid, dilute with water to 1 liter in a volumetric flask. Dilute 25.0 ml of this solution with water to 250 ml in a volumetric flask, and mix.

(*b*) Hydroxylamine hydrochloride solution (100 g $NH_2OH.HCl$ per liter). – Dissolve 100 g of hydroxylamine hydrochloride in ca. 500 ml of warm water, cool, and dilute with water to 1 liter. Prepare fresh as required.

(*c*) Dipyridyl solution (2 g 2,2′-dipyridyl per liter). – Dissolve 0.2 g of 2,2′-dipyridyl in ca. 90 ml of water and 3 ml of HCl (1 + 1), dilute with water to 100 ml, mix, and store in cool, dark place.

(*d*) Sodium acetate buffer solution (272 g $CH_3COONa.3H_2O$ per liter). – Dissolve 272 g of sodium acetate trihydrate in 800 ml of warm water, cool, and dilute with water to 1 liter.

Preparation of calibration curve

Transfer 5, 10, 15, 20, 30, and 40 ml of standard iron solution (reagent *a*) to six 100-ml volumetric flasks. To each flask and to an additonal flask for the blank, add 10.0 ml of hydroxylamine hydrochloride (reagent *b*), mix, and allow to stand for 10.0 min. Then add 5.0 ml of dipyridyl solution (reagent *c*), mix, add 25 ml of sodium acetate buffer solution (reagent *d*), mix, dilute with water to the mark, mix, and allow to stand for five minutes.

Next transfer a suitable portion of the reference solution to an absorption cell having a 1-cm light path. Using a spectrophotometer measure the absorbancy at approximately 522 mμ, compensate or correct for the blank, and plot the values obtained against mg of Fe_2O_3 per 100 ml of solution (Fig.8).

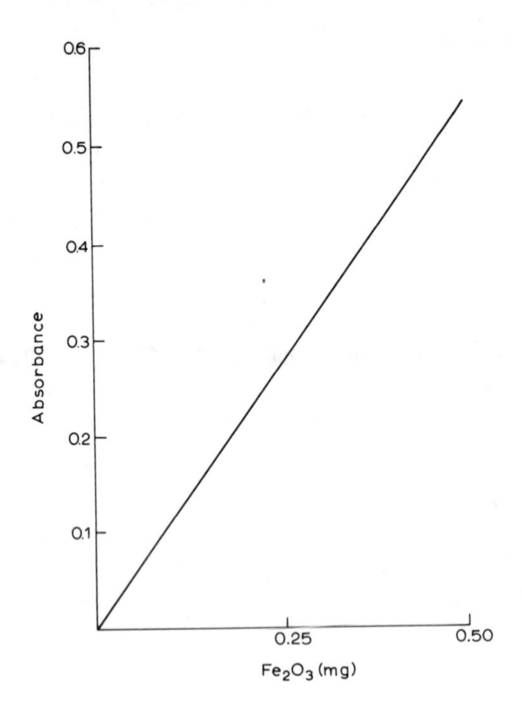

Fig. 8. Calibration curve for ferric iron using a 1-cm cell, 100 ml volume and a wave-length of 522 mμ.

Procedure

From the clear solution of the soluble bases obtained by the use of hydrofluoric and sulphuric acids as described on p. 35, transferred to a graduated flask and diluted to the mark, pipette an aliquot (≡ 0.05–0.50 mg Fe_2O_3) into a 100-ml volumetric flask.

To that flask and to an additional flask for the blank, add 10.0 ml of hydroxylamine hydrochloride (reagent *b*), mix, allow to stand for 10 min to permit the reduction of the iron to the ferrous state, and continue in accordance with "Preparation of calibration curve". Compensate or correct for the blank.

Using the value obtained, read from the calibration curve the number of mg of iron (in terms of Fe_2O_3) present in 100 ml of solution, and calculate the percentage of Total iron (in terms of Fe_2O_3) of the sample as follows:

$$\text{Total ferric oxide \%} = \frac{A - B}{C \times 10}$$

where:

A = mg of iron (determined in terms of ferric oxide) found in the aliquot used;

B = reagent blank correction in mg of iron (determined in terms of ferric oxide);

C = grams of sample represented in the aliquot used.

OTHER METHODS

There are many other procedures which can be applied for the determination of Total iron in silicate analysis but which do not add greatly to those already described; most of them are mainly of academic interest and details of their use should be sought in the original literature.

Gravimetric determination of Total iron after separating by the ion-exchange method

Principle of method

Since iron has a high adsorption coefficient in concentrated hydrochloric acid when an anion-exchange resin is used, this element is retained on the resin while the other members of the ammonium hydroxide group with which it is usually associated, are eluted (see p. 110). The iron is then removed from the resin by means of dilute hydrochloric acid, the iron eluate evaporated to a convenient volume, and the amount of iron determined by gravimetric, volumetric, colorimetric, or photometric methods. The procedure may be applied to the non-volatile residue after decomposition of the material by the use of hydrofluoric acid (see Decomposition by acids, pp. 35–37; also Scheme II).

Preparation of the column

All reagents to be used must be iron-free. Prepare a Dowex 1X4 (50–100 mesh to the linear inch) anion ion-exchange resin column (2 x 7 cm) by passing 100 ml of hydrochloric acid slowly through it; discard the liquid which has passed the column.

If an iron-free resin is not available, free a standard grade resin from iron as follows: allow the resin to stand with occasional swirling in a covered beaker for 2 h with 50 ml of hydrochloric acid, transfer the mixture to a tube similar to a Jones reductor, wash the column carefully with 200 ml of HCl (5 + 95) or more, test washings to be free from iron (in acid solution ferric salts form a red-brown soluble complex with 1 *N* ammonium thiocyanate), and finally pass 100 ml of hydrochloric acid through the column which is now ready for use; discard the hydrochloric acid which has passed the column.

When iron-free reagents are not available, assure the accuracy of the test or the results of the analysis, by running blanks or by checking against comparable samples of known composition.

Procedure

Transfer the clear solution of the soluble salts obtained by the use of hydrofluoric and sulphuric acid as described on p. 35, to a 250-ml beaker, dilute with water to 180 ml, add 20 ml of HCl (1 + 1) and a little bromine water, boil until the excess of bromine is expelled, add a few drops of methyl red indicator (0.2% alcoholic solution), heat just to boiling and, while stirring with a rubber-tipped glass rod, add NH_4OH (1 + 1) dropwise until the red colour of the solution just changes to a light yellow. Boil the solution for not longer than 2 min, add one or more drops of NH_4OH (1 + 1) if the red colour of the indicator reappears, filter at once and rapidly through a wet properly set No.40 Whatman or similar ashless filter paper into a suitable Erlenmeyer flask, wash the impure precipitate 6–8 times with small amounts of a hot neutral ammonium chloride solution (2%); discard filtrate and washings, and reserve precipitate.

Place funnel with contents above the beaker in which the precipitation was made, unfold the paper, place it against the inside of the funnel, return the precipitate to the beaker by washing it off the paper with a strong jet of boiling water, dissolve the precipitate by adding 30 ml of hot HCl (1 + 1), and reserve the washed paper in a platinum crucible.

Char the paper without inflaming, cautiously burn the carbon at a low temperature and under good oxidizing conditions, dissolve the small residue in 10 ml of hot HCl (1 + 1), add the solution and washings to the beaker, and evaporate the combined solutions in a porcelain dish on the steam bath until most of the hydrochloric acid has been expelled.

Take up the residue (chiefly composed of compounds of the elements: iron, aluminium, titanium, phosphorus, vanadium and, if present in the sample, chromium and zirconium) with 20 ml of hydrochloric acid, pour the solution at a rate of 2–5 drops per min via a suitable fritted-glass filter through the prepared Dowex resin column, by which operation iron is adsorbed on the resin, whereas the other elements should be washed down the column and removed from the resin by passing a further 150 ml of hydrochloric acid via the glass filter through the column, allowing to drain well between additions of the washing solution. (If desired, the eluate and washings may be combined and reserved for the determination of the other elements of the ammonia precipitate.)

Release the iron from the resin by washing the column carefully with 200 ml of HCl (5 + 95), collect eluate and washings into a 400-ml beaker, use more HCl (5 + 95) until washings are free from iron (test small amounts of washings by noting change in colour with a freshly prepared 1 *N* aqueous solution of ammonium thiocyanate), transfer the combined eluate and washings to a porcelain dish and evaporate to a volume of 10–20 ml on the steam bath.

To remove any particles of resin that may have passed through the column, dilute the residue with 50–100 ml of water, pour the solution through a suitable fritted-glass filter into a 250-ml beaker, wash dish and filter 10 times dropwise with hot HCl (5 + 95) allowing to drain well between additions of the washing solution, finally suck dry at the pump. Combine and reserve filtrate and washings.

Dilute the combined filtrate and washings with water to a volume of 100–200 ml, add a few drops of methyl red indicator (0.2% alcoholic solution), heat to boiling and, while stirring with a rubber-tipped glass rod, add NH_4OH (1 + 1) dropwise until the red colour of the solution changes to a distinct yellow (with iron alone present, overstepping neutrality is of no consequence), boil the solution for about 2 min, add one or more drops of NH_4OH (1 + 1) if the red colour of the indicator reappears, allow to settle somewhat, filter through a properly set No.40 Whatman or similar ashless filter paper into a suitable Erlenmeyer flask, wash the precipitate 6–8 times with small amounts of a hot neutral ammonium nitrate solution (2%), discard filtrate and washings, and reserve precipitate.

Transfer the precipitate to the paper pouring the solution down a glass rod, detach the film of ferric hydroxide adhering to the beaker by means of a small piece of ashless paper wrapped round the rod, add the small piece to the precipitate, transfer paper and contents to a platinum crucible weighed with well-fitting cover, carefully dry, char the paper without inflaming, burn the carbon at a low temperature and under good oxidizing conditions and, using a Meker burner, heat the covered crucible at 1000–1100°C in an inclined position so that the products of combustion do not envelope the mouth of the crucible.

Allow to cool in a small desiccator over a good desiccant, weigh without delay while still covered, and repeat the ignition and weighing as Fe_2O_3 until constant weight is obtained. (The precipitate must be free from silica derived from the glass and porcelain ware; if there is any doubt, the precipitate should be treated with iron-free hydrofluoric acid, followed by evaporation and ignition before the ferric oxide is weighed.)

CHAPTER 10

Titanium

CONSIDERATION OF METHODS

Titanium is one of the most universally distributed elements. In terrestrial matter it occurs in small concentrations only and comprises about 0.60 per cent of the lithosphere. A number of silicates contain titanium; in most cases the element therein is to be taken as replacing the silicon, in others it seems to play the part of a basic element, e.g., in titanite $CaO.TiO_2.SiO_2$; in some it may enter in both relations as in schormolite $3CaO.(Fe,Ti)_2O_3.3(Si,Ti)O_2$.

In silicate analysis where in the majority of cases the amount of titanium present will not exceed one per cent, the most widely adopted methods are colorimetric. Based on the yellow to amber colour given by titanium in a sulphuric acid solution containing hydrogen peroxide, of sufficient sensitivity, and having the advantage that, where the titanium content of the sample under test may be unknown or proven to be higher than expected, adjustment of the final volume is possible after development of the soluble coloured complex, the method is almost universally used for the colorimetric determination of small amounts of titanium in silicate rocks and minerals. Using this method, it is customary to determine the titanium as titania in an aliquot of filtrate (*15*) of the main portion after the volumetric determination of iron with standardized permanganate in the sulphuric acid solution of the pyrophosphate fusion of the ammonia precipitate as outlined in the general procedure (Scheme I, p. 85), or in aliquot of the clear solution of the soluble bases obtained by direct attack of silicates by hydrofluoric and sulphuric acids as described on p. 35; the presence of manganese derived from the permanganate used in titrating the iron (see p. 100) does no harm.

In preparing solutions for the determination of titanium especially with regard to the wet attack, it should be borne in mind that in the presence of fluorine, volatile titanium tetrafluoride may be formed, also fluorine bleaches the colour of pertitanic acid sulphate; for these reasons fluorine must be removed previously by repeated heating with sulphuric acid. Titanium will hydrolyze if the acidity of its solution is low, however, the presence of sufficient sulphuric acid (a 5–100% v/v solution of H_2SO_4 is to be preferred) not only prevents the formation of metatitanic acid but at the same time insures complete oxidation of the titanium. Where titanium and phosphorus occur together in large quantities, titanium may react with the phosphate ion to form insoluble titanium phosphate; in this case a separation of titanium by precipitating it with, e.g., sodium hydroxide in the presence of a little iron has to be made, followed by redissolving the precipitate in dilute sulphuric acid.

Interference of moderate amounts of elements such as iron, chromium, copper, nickel, and the like, forming coloured compounds but not changing colour on addition of hydrogen peroxide, can be compensated by adding like amounts of the same compounds to the reference solution; large amounts of iron present, their interference can be eliminated by complexing the iron by adding phosphoric acid in like amount to both reference solution and sample solution, after the addition of hydrogen peroxide. (In the method given by Riley, 1958a, a mixed reagent solution consisting of phosphoric acid, hydrogen peroxide, and sulphuric acid, is used.)

The separation of titanium from interfering elements such as vanadium and molybdenum, forming coloured compounds with hydrogen peroxide in dilute sulphuric acid solution, from chromium^{VI}, and also from phosphorus, can easily be made by precipitating titanium with sodium hydroxide. Compounds of the elements fluorine (even in minute amounts) and phosphorus, as well as those of the alkali metals (in large amounts) present in the sample, bleach the colour of the pertitanic acid sulphate; fluorine can be removed by repeated heating with sulphuric acid, and phosphorus by treating the solution with sodium hydroxide, as indicated above.

The bleaching action of small amounts of phosphorus can be compensated by adding like amounts to the reference solution. Large amounts of alkali salts should be removed by treating the solution with ammonium hydroxide and redissolving the titanium precipitate in dilute sulphuric acid, whereas the bleaching effect of moderate amounts of these salts may be reduced and corrected by increasing the sulphuric acid content of their solutions to about 10% v/v, and by adding like amounts of the same alkali salts and sulphuric acid to the reference solution.

SPECTROPHOTOMETRIC DETERMINATIONS

A. Titanium by the hydrogen peroxide method
(see Scheme I, also Scheme II)

Principle of method

Titanium forms a soluble yellow to amber-coloured complex with a moderate excess of hydrogen peroxide in dilute sulphuric acid solution. Photometric measurement is made at approximately 410 mμ.

Concentration range

The recommended concentration range is from 0.15 to 2.5 mg of titanium in 100 ml of solution, using a cell depth of 1 cm.

Stability of colour

The colour is increased by increase of temperature. The colour does not change between 18–27°C, and is stable for hours if reducing agents are absent.

Interfering elements

The elements ordinarily present in filtrate (*15*) of the main portion (Scheme I) or obtained as soluble sulphates after decomposition of the sample by the use of hydrofluoric and sulphuric acids (Scheme II) do not interfere in that the reference solution corrects for the interference of normal constituents. However, traces of fluorine, large amounts of iron, of phosphorus, and of alkali salt, should be absent, also vanadium and molybdenum, giving with hydrogen peroxide in dilute sulphuric acid solution brownish-red and yellow coloured compounds, respectively. If present in disturbing amounts, these elements must be removed by suitable methods before the determination of titanium is attempted (see p. 114). Some interferences have been given in the literature (Sandell, 1959).

Special reagents

(*a*) Standard titanium solution (1 ml ≡ 0.15 mg Ti). – Fuse 0.2550 g of dried titanium dioxide, primary standard grade, in a covered platinum crucible with 5 g of potassium bisulphate, dissolve the cold melt in 200 ml of hot H_2SO_4 (1 + 1), cool to room temperature, transfer the solution to a 1-l volumetric flask, dilute with water to the mark, and mix.

(*b*) Hydrogen peroxide (28–30%, free from fluorine).

Preparation of calibration curve

Transfer 1.0, 2.0, 5.0, 7.0, 9.0 and 10.0 ml of titanium solution (reagent *a*) to six 100-ml volumetric flasks. To each flask and to an additonal flask for the reference solution, add enough H_2SO_4 (1 + 9) to make a total of 80 ml, and also 2 drops of H_2O_2 (reagent *b*). Dilute to the mark with H_2SO_4 (1 + 9), and mix.

Transfer a suitable portion of the reference solution to an absorption cell having a 1-cm light path and adjust the photometer to the initial setting, using a light band centered at approximately 410 mμ.

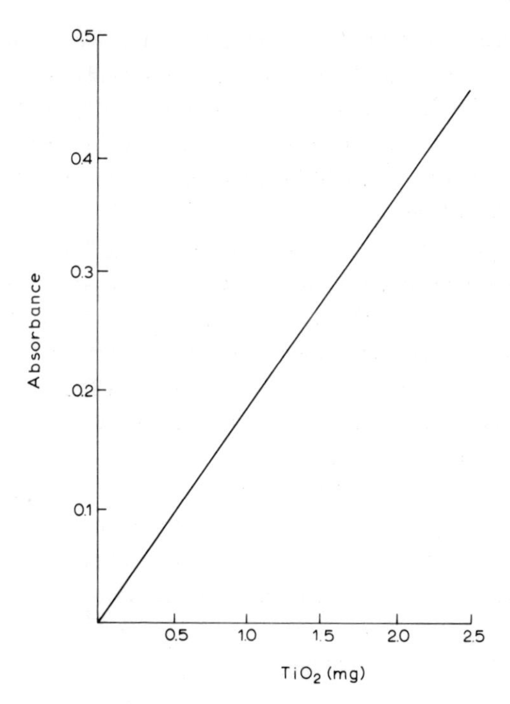

Fig.9. Calibration curve for titanium (peroxide complex) using a 2-cm cell, 100 ml volume and a wave-length of 400 mμ.

Maintaining this photometer adjustment, take the photometric readings of the calibration solutions. Plot the values obtained against mg of Ti per 100 ml of solution (Fig.9).

Procedure

Following this procedure, also make a blank determination (B), using the same amounts of all reagents.

Using filtrate (*15*) of the main portion, that has served for the titration of Total iron (see Scheme I), evaporate to less than 100 ml, cool, transfer the concentrated filtrate to a 100-ml volumetric flask, dilute to the mark with water, and mix.

Otherwise, using the clear solution of the bases, obtained by direct attack by hydrofluoric and sulphuric acids as described on p. 35 (and freed, when necessary, from interfering contaminations by suitable methods), add enough sulphuric acid to make a total of 10 ml, cool, transfer the solution to a 100-ml volumetric flask, dilute to the mark with water, and mix.

Transfer two similar aliquots ($\equiv 0.15-1.5$ mg Ti) of the solution obtained by one of the methods given, to two 100-ml volumetric flasks, dilute with H_2SO_4 (1 + 9) to the mark, and mix. Reserve one of the diluted aliquots for use as a reference solution (R); add to the other one 2 drops of H_2O_2 (reagent *b*), and mix (S).

Next transfer a suitable portion of the reference solution (R) to the absorption cell used before, and take the photometric reading of the solution (S) as described in "Preparation of calibration curve".

Convert the photometric readings of the final sample and blank solutions to mg of titanium by means of the calibration curve. Calculate the percentage of titanium (or of titania) as follows:

$$\text{Titanium } \% = \frac{A - B}{C \times 10}$$

$$\text{Titania } \% = \frac{(A - B) \times 1.67}{C \times 10}$$

where:

A = mg of titanium found in the aliquot used;

B = reagent blank correction in mg of titanium;

C = grams of sample represented in the aliquot used.

(Note: after the colour comparison the peroxidized sample solution should be treated with 2 drops of hydrofluoric acid to discharge the titanium colour and so confirm the absence of vanadium.)

B. Titanium by the tiron method
(see Scheme I, also Scheme II)

Principle of method

Titanium forms a soluble yellow-coloured complex with tiron (disodium 1,2-dihydroxybenzene 3,5-disulphonate) in a buffered solution. Photometric measurement is made at approximately 430 mμ.

Concentration range

The method is confined to minute amounts of titanium; the recommended concentration range is from 0.01 to 0.10 mg of TiO_2 in 30 ml of solution, using a cell depths of 1 cm.

Stability of colour

The colour of the titanium complex develops within 5 min, and then starts to fade slowly.

Interfering elements

The elements ordinarily present in filtrate (*15*) of the main portion (Scheme I), or else obtained as soluble sulphates after decomposition of the sample by the use of hydrofluoric and sulphuric acids (Scheme II) do not interfere in that the reference solution corrects for the interference of normal constituents. Ferric iron interferes, forming with tiron a highly coloured purple complex. When present in small amounts ($<$5 mg Fe per 100 ml of solution) the ferric complex can be largely decolorized by reducing the iron to the ferro state by addition of sodium dithionite after the titanium complex has been formed. When present in large amounts (5–10 mg Fe per 100 ml of solution) and excessive quantities of sodium dithionite would be required, iron must be eliminated previously either by complexing with EDTA, (p. 122), or by adsorption on an anion

ion-exchange resin in hydrochloric acid solution (p. 110); or ferric chloride by extraction with ether (p. 97).

Special reagents

(*a*) Standard titanium solution (1 ml ≡ 0.02 mg TiO_2). – Transfer a 10.0-ml aliquot of the previously given standard titanium solution to a 100-ml volumetric flask, dilute with H_2SO_4 (1 + 99) to the mark, and mix. Prepare fresh as needed.

(*b*) Tiron (disodium 1,2-dihydroxybenzene 3,5-disulphonate). – Keep in a dark-coloured, glass-stoppered bottle.

(*c*) Ammonium acetate buffer solution. – Dissolve 40 g of CH_3COONH_4 in 500 ml of water, add 15 ml of CH_3COOH, and dilute with water to 1 liter.

(*d*) Sodium dithionite ($Na_2S_2O_4.2H_2O$). – Keep in well-stoppered bottle.

Preparation of calibration curve

Transfer 0.5, 1.0, 2.0, 3.0, 4.0, and 5.0 ml of titanium solution (reagent *a*) to six dry 100-ml beakers. To each beaker and to an additional beaker for the blank, add H_2SO_4 (1 + 99) to make a total of 5.0 ml, 100 mg of tiron (reagent *b*), and 25.0 ml of buffer solution (reagent *c*). Mix, add 10 mg of sodium dithionite (reagent *d*), and stir carefully until all has dissolved.

Immediately transfer a suitable portion of the reference solution to an absorption cell having a 1-cm light path. Using a spectrophotometer measure the absorbancy at approximately 430 mμ, compensate or correct for the blank. Plot the values obtained against mg of TiO_2 per 30 ml of solution (Fig.10).

Procedure

Dilute filtrate (*15*) of the main portion (Scheme I), or else the clear solution of the soluble bases obtained by direct attack of the

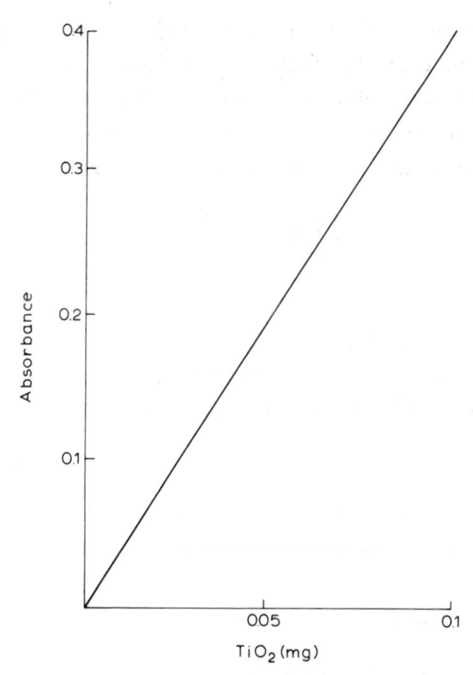

Fig.10. Calibration for titanium (tiron complex) using a 1-cm cell, 30 ml volume and a wave-length of 4.30 mμ.

sample by hydrofluoric and sulphuric acids as described on p. 35, freed from harmful contaminants (see p. 114) and containing 10% v/v of sulphuric acid, to a suitable volume in a volumetric flask with H_2SO_4 (1 + 9), and mix. Pipette 10.0 ml of the diluted solution into a 100-ml volumetric flask, dilute with water to the mark, and mix.

Transfer a 5-ml aliquot or less (≡ no more than 0.10 mg TiO_2) to a dry 100-ml beaker, add H_2SO_4 (1 + 99) to make a total of 5.0 ml, 100 mg of tiron (reagent *b*), and 25.0 ml of buffer solution (reagent *c*), and mix. To an additional 100-ml beaker for the blank add the same amounts of all reagents that are present in the unknown. Just prior to the measurement of the absorbancy, add to both beakers 10 mg of sodium dithionite (reagent *d*), and stir carefully until the salt has dissolved. Continue in accordance with "Preparation of calibration curve". Compensate or correct for the blank.

Using the value obtained, read from the calibration curve the number of mg of titania present in the aliquot. Calculate the percentage of titanium (in terms of titania) of the sample as follows:

$$\text{Titania } \% = \frac{A - B}{C \times 10}$$

where:

A = mg of titania found in the aliquot used;
B = reagent blank correction in mg of titania;
C = grams of sample represented in the aliquot used.

C. Titanium by the EDTA-tiron method
(see Scheme I, also Scheme II)

Principle of method

Titanium forms a soluble yellow-coloured complex with tiron (disodium 1,2-dihydroxybenzene 3,5-disulphonate) in a buffered solution. Iron, if present in quantities <10 mg per 50 ml of solution, is held as a colourless ferrous-EDTA complex. Photometric measurement is made at approximately 430 mμ.

Concentration range

The method virtually designed for the direct determination of titanium in meteorities, is suitable for the determination of minute amounts of titanium in the presence of large quantities of iron. The recommended concentration range is from 0.01 to 0.10 mg of TiO_2 in the presence of as much as 10.0 mg of Fe per 50 ml of solution, using a cell depth of 1 cm.

Stability of colour

The full colour of the titanium-tiron complex develops within 5 min. Photometric measurement should be made within 15 min after adding the sodium dithionite, because decomposition of this salt may lead to turbidity on standing.

Interfering elements

The elements ordinary present in filtrate (*15*) of the main portion (Scheme I), or else obtained as soluble sulphates after decomposition of the sample by the use of hydrofluoric and sulphuric acids (Scheme II) do not interfere in that the reference solution corrects for the interference of normal constituents. Ferric iron interferes, forming with tiron a highly-coloured purple complex. When present in amounts <10 mg of Fe per 50 ml of H_2SO_4 (1 + 99) solution and complexed with EDTA, the ferric complex can be decolorized by reducing the iron to the ferro state by treating with sodium dithionite *after* the addition of the sodium acetate buffer to a final pH of 5.6, so also avoiding interference in the formation of the yellow-coloured titanium-tiron complex. Excessive iron must be eliminated previously by adsorption on an anion ion-exchange resin in hydrochloric acid solution (p.110); as ferric chloride by extraction with ether (p. 97); or in H_2SO_4 (1 + 99) solution (together with chromium, nickel, and copper) by electrolysis with mercury cathode.

Special reagents

(*a*) Standard titanium solution (1 ml ≡ 0.02 mg TiO_2). – Prepare as in *B,a* (p.120).

(*b*) Tiron solution (20 g Tiron per liter). – Dissolve 2 g of 1,2-dihydroxybenzene 3,5-disulphonic acid in 100 ml of water. Prepare fresh as needed.

(*c*) EDTA solution (2 g Na_2EDTA per liter). – Dissolve 0.2 g of disodium dihydrogen ethylenediamine tetraacetate in 100 ml of water. Prepare fresh as needed.

(*d*) Acetate buffer solution (136 g $NaC_2H_3O_2.3H_2O$ per liter). – Dissolve 27.2 g of sodium acetate trihydrate in 200 ml of water. Prepare fresh as needed.

(*e*) Sodium dithionite ($Na_2S_2O_4.2H_2O$). – Keep in well-stoppered bottle.

Preparation of calibration curve

Transfer 0.5, 1.0, 2.0, 2.0, 4.0, and 5.0 ml of titanium solution (reagent *a*) to six 50-ml volumetric flasks. To each flask and to an additional flask for the blank, add H_2SO_4 (1 + 99) to make a total of 5.0 ml, 5.0 ml of tiron solution (reagent *b*), 5.0 ml of EDTA solution (reagent *c*), 10.0 ml of buffer solution (reagent *d*), dilute to the mark with water, stopper the flask, and mix. Just prior to the measurement add ca 50 mg of sodium dithionite (reagent *e*), stopper, and shake until all has dissolved.

Immediately transfer a suitable portion of the reference solution to an absorption cell having a 1-cm light path. Using a spectrophotometer measure the absorbancy at approximately 430 mμ, compensate or correct for the blank. Plot the values obtained against mg of TiO_2 per 50 ml of solution.

Procedure

Dilute filtrate (*15*) of the main portion (Scheme I), or else the clear solution of the soluble bases obtained by direct attack by hydrofluoric and sulphuric acids as described on p. 35, freed from harmful contaminants (see p. 114) and containing 10% v/v of sulphuric acid, to a suitable volume in a volumetric flask with H_2SO_4 (1 + 99), and mix. Pipette 10.0 ml of the diluted solution to a 100-ml volumetric flask, dilute with water to the mark, and mix.

Transfer an aliquot (≡ 0.01–0.10 mg TiO_2) to a 50-ml volumetric flask, add H_2SO_4 (1 + 99) to make a total of 5.0 ml, 5.0 ml of tiron solution (reagent *b*), 5.0 ml of EDTA solution (reagent *c*), 10.0 ml of buffer solution (reagent *d*), dilute to the mark with water, stopper the flask, and mix. To an additional 50-ml volumetric flask for the blank add the same amounts of all reagents that are present in the unknown. Just prior to the measurement of the absorbancy, add to both flasks 50 mg of sodium dithionite (reagent *e*), stopper, and shake until all has dissolved. Continue in accordance with "Preparation of calibration curve". Compensate or correct for the blank.

Using the value obtained, read from the calibration curve the number of mg of titania present in the aliquot. Calculate the percentage of titanium (in terms of titania) of the sample as follows:

$$\text{Titania \%} = \frac{A - B}{C \times 10}$$

where:
A = mg of titania found in the aliquot used;
B = reagent blank correction in mg of titania;
C = grams of sample represented in the aliquot used.

OTHER METHODS

There are some other procedures which can be applied for the colorimetric determination of small amounts of titanium in silicate analysis but which do not add greatly to the spectrophotometric methods already described. Among them are determinations based on the coloured complex which titanium gives: with tiron and sodium dithionite in a buffered solution (Yoe and Armstrong, 1947), see p. 119; with sodium alizarine sulphonate and stannous chloride in HCl solution (Goto et al., 1957); in the presence of small amounts of iron, with tiron, thioglycollic acid and sodium dithionite (Rigg and Wagenbauer, 1961); or in conjunction with EDTA instead of thioglycollic acid (Easton and Greenland, 1963) when large amounts of iron are present, see p. 122; with 3,6-dichlorochromotropic acid and ascorbic acid (Klassova and Leonova, 1964).

CHAPTER 11

Aluminium

CONSIDERATION OF METHODS

Next to oxygen and silicon, aluminium is the most abundant element; it comprises about 8.0 per cent of the lithosphere. It does not occur native, and it always occurs in nature in the trivalent state. Aluminium is found chiefly in silicates, such as the feldspars (orthoclase, albite, labradorite, etc.), micas (muscovite, lepidolite, etc.), and clays (kaolinite, halloysite, etc.) the last being decomposition products of certain aliminous minerals. It is an essential component of the bauxites (mixtures of varying amounts of a colloidal form of $Al_2O_3.H_2O$, associated with iron sesquioxide, silica, phosphoric acid, etc., as common impurities), it occurs as basic hydrous sulphate in alunite, as fluoride in cryolite, as the oxide carborundum, as phosphate in lazurite, and in limited amounts in many other compounds.

The most used method for the separation and determination of aluminium in silicate rocks and minerals is the oxide method, in which aluminium is precipitated by ammonium hydroxide and weighed as the oxide, Al_2O_3. Notwithstanding the fact that members of the acid and hydrogen sulphide groups can be removed beforehand, so many other elements still in solution may accompany the aluminium completely or in part, that the precipitation by ammonium hydroxide generally serves only as a preliminary separation. However, using that procedure, the determination of aluminium can be completed by ignition and one weighing of all as oxides and determination of the weights of all accompanying oxides to obtain the weight of aluminium, as Al_2O_3, by difference. Moreover, its determination can usually be simplified by determining some of the components in groups, e.g., by concentrating or diluting filtrate (*15*) (Scheme I) to a definite volume and dividing the adjusted filtrate into three suitable portions. One portion may be treated with ammonium hydroxide and the precipitate ignited to obtain the total weight of the members of the ammonium hydroxide group as oxides (Al_2O_3, Fe_2O_3, TiO_2, ZrO_2, P_2O_5, V_2O_5, and more or less of the Cr_2O_3 if present in the sample), the second portion in strong acid solution should be treated with cupferron and the precipitate ignited to obtain the total weight of the elements precipitated by cupferron as oxides (Fe_2O_3, TiO_2, ZrO_2, V_2O_3), whereas the third portion should be used for the determination of P_2O_5 by means of the molybdate-magnesia method, to obtain the weight of the Al_2O_3 + more or less of the Cr_2O_3 by difference. As a rule so little chromium is present in the sample that its amount can be ignored in routine analyses; when desired, the quantity of chromium can be determined colorimetrically by the sodium peroxide fusion method and matching the colour of its solution with that of a standard solution. The determination of Al_2O_3 by difference, by using three aliquot portions of the filtrate (*15*) (Scheme I) is quite satisfactory if the sample contains elements (besides aluminium and phosphorus) that are precipitated by ammonium hydroxide as well as by cupferron (amino-nitroso-phenylhydroxylamine) being a very useful precipitant for quantitative procedures introduced into analysis by Baudisch (1909).

Oxine (8-hydroxyquinoline) forming compounds with aluminium and several other elements in feebly acid or alkaline solutions, was proposed by Berg (1927); showing wide differences in solubility the compounds in feebly acid solutions can be precipitated and separated by buffering and controlling the pH. However, cupferron and oxine methods for aluminium, being group precipitations, involve standard separations, the determination of aluminium being completed by gravimetric, volumetric, or colorimetric methods.

As standard separations for the isolation of aluminium from complex mixtures may be mentioned its separation from members of the hydrogen sulphide group by hydrogen sulphide in acid solution; together with phosphorus, vanadium, and the like, from members of the ammonium hydroxide group by precipitating iron, titanium, zirconium, and the rare earths, with sodium hydroxide; from iron (preferably after a preliminary removal of the members of the hydrogen sulphide group) by precipitating the iron with ammonium sulphide in an ammoniacal solution containing tartrate thus keeping the other members of the ammonium hydroxide group in solution, by removing the iron with ether in hydrochloric acid solution (p. 97), by ion-exchange separation, or by other selective methods (see p. 110); together with titanium, vanadium, zirconium, and the like, from iron, chromium, manganese, nickel, cobalt, copper, and zinc, by electrolysis with the mercury cathode in dilute sulphuric acid solution.

DETERMINATION OF ALUMINA IN SILICATE ROCKS

Alumina by "difference"

The most used method for the determination of alumina in silicate rocks and minerals is that as outlined in Scheme I in which aluminium together with all of any residual SiO_2, P_2O_5, TiO_2, ZrO_2, or other members of the ammonium hydroxide group that may be present in the sample, is first separated from the alkalies, alkaline earths, magnesium, manganese, and elements such as nickel and zinc, by precipitation twice with ammonium hydroxide in the presence of ammonium chloride (see Total iron, p. 98–99) followed by ignition and one weighing of all as Impure Mixed oxides (*13*), and determination of the weights of all accompanying oxides one after another by methods of separation and determination mentioned at various places throughout the book, and so to obtain the weight of aluminium, as Al_2O_3, by "difference". In silicate analyses the quantities of the accompanying oxides usually being relatively small, the oxide method yields results that are quite satisfactory.

$$\text{Alumina \%} = \frac{(A - B)}{C} \times 100$$

where:
A = grams of total weight of the Impure Mixed oxides (*13*);
B = grams of total weight of all accompanying oxides;
C = grams of sample used.

VOLUMETRIC DETERMINATION

Alumina by the EDTA method

Principle of method

Aluminium may be determined volumetrically by first complexing it with EDTA (ethylenediamine tetraacetic acid), removing the excess, and then liberating the EDTA from the complex by addition of sodium fluoride (Kinnunen and Wennerstrand, 1957). The excess as well as the liberated EDTA are titrated at pH 5–6 with zinc acetate. A suitable indicator is an aqueous solution of xylenol orange (Korbl and Pribil, 1956). Since both the formation and decomposition of the EDTA-aluminium complex is slow at room temperature, the solution is heated to boiling. Iron must be absent and should be removed beforehand by one of the methods given on p. 128. Titanium is complexed by the addition of phosphate.

(Another volumetric procedure for the determination of aluminium after the removal of iron, has been proposed (Milner and Woodhead, 1954); here the excess of EDTA added is titrated with ferric chloride in a solution containing 3 g of ammonium acetate and neutralized with ammonium hydroxide to a pH 6.5, using the iron salicylate colour to indicate the end point.)

Special reagents

(*a*) EDTA solution (7.5 g Na_2EDTA per 1). – Dissolve 7.5 g of disodium dihydrogen ethylenediamine tetraacetate in 900 ml of double distilled water, transfer the solution to a 1-l volumetric flask, dilute to the mark, and mix. Transfer the solution to a well-stoppered polyethylene bottle.

(*b*) Sodium acetate buffer solution (100 g $CH_3COONa.3H_2O$ per liter). – Dissolve 10 g of sodium acetate trihydrate in water and dilute to 100 ml.

(*c*) Methyl red indicator paper, pH range 4.2–6.3.

(*d*) Monosodium phosphate solution (100 g $NaH_2PO_4.2H_2O$ per liter). – Dissolve 10 g of monosodium phosphate dihydrate in water and dilute to 100 ml.

(*e*) Xylenol orange indicator solution (2 g per l). – Dissolve 0.2 g of xylenol orange in 100 ml of water.

(*f*) Zinc acetate solution [4.4 g $(CH_3COO)_2Zn.2H_2O$ per l]. – Dry zinc acetate dihydrate at 100°C and cool in desiccator to constant weight. Dissolve 4.4 g of the dried salt in water, transfer the solution to a 1-l volumetric flask, dilute with water to the mark, and mix.

(*g*) Ammonium fluoride (NH_4F, solid).

(*h*) Standard aluminium solution (1 ml ≡ 1 mg Al_2O_3). – Dissolve 0.5291 g of purest obtainable aluminium foil in a covered 500-ml beaker in the minimum quantity of NaOH (10%) on the steam bath until solution is practically complete, and dilute to ca. 300 ml with water Carefully add H_2SO_4 (1 + 9), drop by drop, until the colour of a methyl red indicator paper (reagent *c*) changes to a distinct red, cover the beaker, and boil until all has dissolved. Cool to room temperature, transfer the solution to a 1-l volumetric flask, dilute with water to the mark, and mix.

Procedure

Measure an aliquot (≡ ca. 50 mg Al_2O_3) of the sample solution, freed from iron by one of the methods given on p. 128, into a 300-ml conical flask, and add an excess of 1 ml of EDTA solution (reagent *a*) for each 5 mg of Al_2O_3-equivalent present in the aliquot under test. Dilute to 150 ml, heat the mixture to boiling, add sufficient sodium acetate buffer solution (reagent *b*) and use methyl red indicator paper (reagent *c*) to adjust the pH of the mixture to about 5–6; if titanium is present, add suficient (e.g., 1–2 ml) of mono-sodium phosphate solution (reagent *d*) to complex the titanium.

Cool the mixture to room temperature, dilute to ca. 200 ml, add a few drops of xylenol orange indicator (reagent *e*), and titrate the excess of EDTA present with zinc acetate solution (reagent *f*); the end point is reached when the colour of the mixture changes from yellow to purplish-pink.

To the purplish-pink coloured mixture add 0.5 g of ammonium fluoride (reagent *g*), heat to boiling to decompose the EDTA-aluminium complex, cool to room temperature, and titrate the liberated EDTA with zinc acetate solution (reagent *f*). The volume of the zinc acetate solution used in the last titration allows the aluminium present in the measured aliquot of the sample to be calculated as Al_2O_3, when the zinc acetate solution (reagent *f*) is standardized against the standard aluminium solution (reagent *h*) which has been carried through the method.

SPECTROPHOTOMETRIC DETERMINATION

Alumina by the 8-hydroxyquinoline method

Extraction by chloroform and measurement of the aluminium 8-hydroxyquinolate compound from an aliquot of the sample solution, is not only of particular use where the aluminium content of the sample is less than 5% and only small amounts of interfering elements such as iron, titanium, zirconium, and phosphorus are present, but also in semi-micro analysis (Riley, 1958a). Using this procedure, interference by small amounts (<1 mg) of iron can be prevented by reducing the iron to the ferrous state by means of a reductant and then complexing it in feebly acid solution with e.g., dipyridyl, whereas corrections can be made when other interfering elements are present. After the removal of elements that are not wanted in solution, this procedure may be applied to an aliquot of the nonvolatile residue left after decomposition of the sample by the use of hydrofluoric and sulphuric acids (Scheme II).

(The use of pyrocatechol violet for the spectrophotometric determination of aluminium on a semi-micro scale has been described by Wilson and Sergeant, 1963a.)

Principle of method

After the removal, or complexing, of interfering elements, small amounts of aluminium in faintly acid solution are separated by reaction with 8-hydroxyquinoline and extraction with chloroform. The colour intensity of the resulting yellow solution is determined photometrically at approximately 410 mμ.

Concentration range

The recommended concentration range is from 0.01 to 0.10 mg of Al_2O_3 in 25 ml of solution, using a cell depth of 1 cm.

Stability of colour

Extraction of the light-sensitive aluminium 8-hydroxyquinolate should be made protected from direct sunlight. Readings should be made immediately after diluting to the proper volume.

Interfering elements

Provision must be made for preventing, or compensating for, interference from elements mentioned on p. 131. Interference of traces of fluorides, derived from the HF-H_2SO_4 decomposition of the sample can be prevented by reaction with beryllium sulphate present in the complexing solution (reagent *b*). For titanium present in the maximum limit, a correction of 0.8 times the percentage of titanium present in the sample is required to be deducted (Davis, 1961).

Special apparatus

A mechanical shaker assists in standardizing extraction of the aluminium 8-hydroxyquinolate compound with chloroform. If the shaker is capable of various speeds, a mark should be fixed on the dial to which the control is set each time an extraction is made. If this type is not available, the standardizing extractions should be done in an ordinary separatory funnel with each set of determinations made on samples.

Special reagents

(*a*) Standard aluminium solution (1 ml ≡ 0.01 mg Al_2O_3). – Transfer 10.0 ml of the standard aluminium solution (p. 130 reagent *h*) to a 1-l volumetric flask, dilute with water to the mark, and mix.

(*b*) Complexing solution. – Dissolve and transfer in order to a 1-l volumetric flask: 10 g of hydroxylamine hydrochloride in 40 ml of water, 3.4 g of sodium acetate trihydrate in 50 ml of water, 0.40 g of 2, 2′-dipyridyl in 20 ml of HCl (1.5 + 98.5), and 0.40 g of beryllium sulphate in 10 ml of water. Mix, dilute with water to the mark, and mix.

Caution: Compounds of beryllium offer a serious health hazard; precautions must be taken against dusting and spraying.

(*c*) 8-Hydroxyquinoline solution (5 g per liter). – Dissolve 1.25 g of 8-hydroxyquinoline in 250 ml of chloroform. Prepare solution freshly every two weeks and store in amber glass, glass-stoppered bottle in a cool place.

Preparation of calibration curve

Protected from direct sunlight prepare the reference solutions as follows: transfer 1, 2, 3, 5, 7, 9, and 10 ml of aluminium solution (reagent *a*) separately to a mechanical shaker (or ordinary separatory funnel), adjust the volume with water to 10.0 ml, use 10.0 ml of water for the blank, add 10.0 ml of complexing solution (reagent *b*), mix, and allow to stand for 5 min.

Next add 20.0 ml of 8-hydroxyquinoline solution (reagent *c*), stopper the separatory apparatus, shake the solution for 5 min, let settle, and draw off the chloroform layer through a small funnel, containing a small plug of ashless filter paper, into a 25-ml volumet-

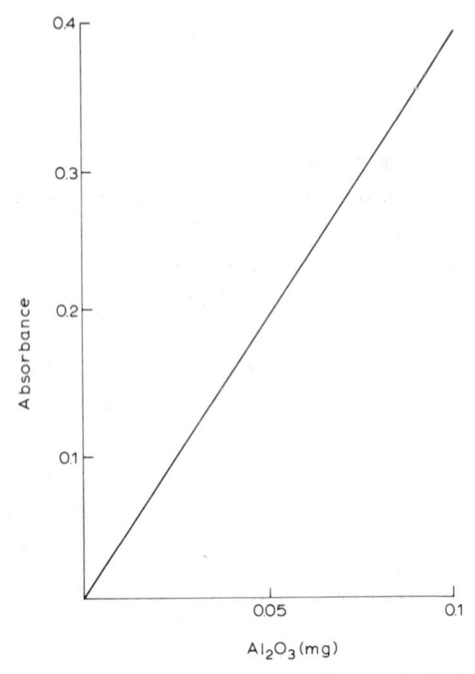

Fig.11. Calibration curve for aluminium using a 1-cm cell, 25 ml volume and a wave-length of 410 mμ.

ric flask. Repeat the extraction with 2–3 ml of chloroform, let settle and draw, and collect the chloroform layer into the volumetric flask before. Dilute the extract with chloroform to the mark, and mix.

Immediately transfer a suitable portion of the reference solution to an absorption cell having a 1-cm light path. Using a spectrophotometer, measure the absorbancy at approximately 410 mμ, compensate or correct for the blank, and plot the values obtained against mg of Al_2O_3 per 25 ml of solution (Fig.11).

Procedure

Dilute the clear solution of the soluble bases obtained by direct attack of the sample by hydrofluoric and sulphuric acids as described on p. 35, freed from harmful contaminants (see p. 131), to a suitable volume in a volumetric flask.

Pipette an aliquot (≡ 0.01–0.10 mg Al_2O_3) of the diluted solution into the separatory apparatus, adjust the volume with water to 10.0 ml, use 10.0 ml of water for the blank, add 10.0 ml of complexing solution (reagent *b*), mix, allow to stand for 5 min, and continue in accordance with "Preparation of calibration curve". Compensate or correct for the blank.

Using the value obtained, read from the calibration curve the number of mg of aluminium (in terms of Al_2O_3) present in the aliquot. Calculate the percentage of alumina of the sample as follows:

$$Al_2O_3\ \% = \frac{A-B}{C \times 10}$$

where:

A = mg of alumina found in the aliquot used;
B = reagent blank correction in mg of alumina;
C = grams of sample represented in the aliquot used.

CHAPTER 12

Calcium

CONSIDERATION OF METHODS

Calcium comprises over 3.6 per cent of the lithosphere, in which it is combined. It occurs as polysilicate in plagioclase feldspars (i.e., in anorthite and andesine), as metasilicate (Ca, Mg, etc., silicate) in several members of the pyroxene and amphibole groups (i.e., in wollastonite and hornblende), as orthosilicate in some garnets and epidotes (i.e., in grossularite and zoisite), as titano-silicate in titanite, and is also an essential constituent of many hydrous silicates.

As a rule calcium is determined gravimetrically by the oxide or sulphate methods (p. 142 and 144), or volumetrically by the permanganate or EDTA methods (p. 144 and 145). In almost all silicate rocks the most used method for the separation and determination of appreciable amounts of calcium, in absence of excessive amounts of phosphorus or magnesium, is that as outlined in Scheme I, in which calcium, together with all of any strontium, is separated by double precipitation from filtrate (*21*) as the oxalate after preliminary separation with ammonium sulphide of all the other elements save strontium, magnesium, barium, the alkalies, and a small part (usually not over 0.2 mg) of manganese and the major part of the platinum from the vessels, which escape precipitation as sulphide. If originally present in the sample, a preliminary treatment of filtrate (*12*) with ammonium sulphide is demanded to remove constituents such as nickel, cobalt, copper, zinc, and also the major part of manganese, that may interfere or that are more or less carried down by members of the succeeding groups, the minor part of that present being caught with the Ammonium oxalate group, the major part with the Ammonium phosphate group.

The precipitation of calcium as oxalate is never quantitative; the small amount (usually not over 0.5 mg) of calcium that always escapes double precipitation in feebly alkaline solution with ammonium oxalate, will be precipitated as phosphate along with the Ammonium phosphate group (see p. 149) and raise, chiefly as $Ca_3(PO_4)_2$, the magnesium pyrophosphate value of the ignited precipitate, thus causing a minus error in the determination of calcium oxide, and a plus error in the subsequent determination of magnesia. On the other hand, in precipitating calcium in feebly alkaline solution with oxalate, traces of undissociated magnesium oxalate are not only occluded by the calcium oxalate as formed to an amount directly proportional to the concentration, but are also adsorbed by the calcium oxalate to an amount about proportional to the time that passes before filtration, both errors causing a plus error in the determination of calcium oxide and a minus error in the subsequent determination of magnesia. Because as a rule, the loss in calcium in the determination of calcium oxide and the loss in magnesium in the determination of magnesia, tend to compensate for the usual plus and minus errors that are made in determination of these elements, in ordinary analyses no attempt is made to correct for the coprecipitated element in question; the difference between the arithmetical figures of the errors is usually too small to make an appreciable error in either determination. As the recovery of small amounts of magnesium from calcium oxide is more laborious than the recovery of small amounts of calcium from magnesium pyrophosphate, in accurate analyses the most satisfactory procedure lies in creating conditions to effect almost complete separation of the magnesium, and in recovering of the calcium therein as detailed on p.153; the amount of calcium thus recovered, calculated to CaO, must be added to the gross weight of the ignited oxalates (p. 155).

If originally present in the sample and the separation of the Ammonium sulphide group has been omitted, besides a small amount of coprecipitated magnesium, the ignited oxalates can be expected to contain practically all of any strontium, as SrO, usually contaminated by a small part of the manganese, nickel, and cobalt.

When strontium is to be sought (inspection of the ignited oxalates by spectrochemical test and visual examination of emission spectrum!), the mixed oxides resulting from ignition of the oxalates must be weighed, converted to nitrates, strontium isolated by treatment with ether-alcohol, converted to sulphate, weighed as $SrSO_4$ and, calculated to SrO, subtracted from the gross weight of the ignited oxalates (p. 143). However, in almost all silicate rocks the quantity of strontium to be present in a portion as that taken for the General procedure, is usually too small to justify a laborious ether-alcohol separation.

When the sample contains manganese in appreciable amounts, and no preliminary ammonium sulphide separation has been made, the minor part of the element will be caught with the Ammonium oxalate group and found as Mn_3O_4 in the impure CaO residue (mostly indicated by a yellow to brown discoloration of the calcium oxide), thus raising the calcium oxide result. In most silicate rocks, however, the manganese content is too low (usually below 0.2%) to make an appreciable error in the calcium oxide determination, and the step for its removal may be omitted. The same would hold for nickel and cobalt. Because of the very small amounts in which these last would occur in silicate rocks, it is customary to disregard their oxides in the ignited oxalates (p. 142).

Although under special conditions superior to the oxalate method, the precipitation of calcium by sulphuric acid and alcohol from a slightly acid hydrochloric or nitric acid solution, is as a rule more used as an initial or group separation in which the three alkaline earths may be separated from other elements. When calcium is to be determined, resolution of the ignited and weighed precipitate and further treatments of the sulphates will become necessary when strontium and barium are present.

Being caught on a fritted-glass filter, the calcium oxalate obtained according to the General procedure (p. 141), if free from strontium (magnesium or barium) oxalate and adhering ammonium oxalate solution, and dissolved in dilute sulphuric acid, may be titrated with a standard potassium permanganate solution; the volume of permanganate solution required must be corrected for that used by the reagents alone. From the corrected amount of potassium permanganate used for titration of the calcium oxalate, the percentage of CaO in the sample may be calculated.

Another volumetric procedure for the determination of calcium oxide is the EDTA method, in which the ignited oxalates (p. 142) are dissolved in a minimum of dilute hydrochloric acid and calcium in a fairly strong (pH 12) alkaline solution is titrated with Na_2EDTA using ethylenedinitrilo tetraacetate as indicator. The volume of Na_2EDTA solution used in the titration allows the CaO present in the sample to be calculated. Using the EDTA method in combination with micro-burettes, very small quantities of calcium can be determined accurately (e.g., the small amounts of calcium that escape precipitation as oxalate and will be caught with the Ammonium phosphate group, see p. 149).

Large amounts of salts obstruct the indicator used (Patton and Reader, 1956). A fairly strong alkaline solution restricts interference of magnesium by precipitating it as the hydroxide. Manganese, if present in disturbing amounts, should be previously removed by treating the filtrate from the ammonium hydroxide group with ammonium sulphide and filtering as described on p. 140, or less satisfactory by catching it in the ammonia precipitate by the use of bromine.

DETERMINATION OF CALCIUM OXIDE IN SILICATE ROCKS IN ABSENCE OF EXCESSIVE AMOUNTS OF MAGNESIUM AND PHOSPHORUS

A. After precipitation as oxalate
(Calcium content > magnesium content)

If originally present in the sample and the separation of the Acid and Ammonium hydroxide groups have been properly made, besides elements of the two succeeding groups the filtrate (*12*) from the Ammonium hydroxide group can be expected to contain all of any manganese and nickel, most of the cobalt and zinc, and a small part of the vanadium. At this point the procedure should be varied, depending on whether or not an ammonium sulphide separation is necessary.

(*a*) When the sample contains copper, cobalt, nickel, zinc, or manganese, in determinable or disturbing amounts a preliminary ammonium sulphide separation is required. In that case, dilute the combined filtrate and washings (*12*) from the Ammonium hydroxide group in a 200-ml Erlenmeyer flask to 100 ml, cool to room temperature, add 2 ml of ammonium hydroxide, pass in a stream of washed hydrogen sulphide until the solution is satured, add again 2 ml of ammonium hydroxide, fill the flask to the neck with recently boiled and cooled water, stopper, and allow the solution to stand overnight. Next, filter the solution uninterruptedly through a medium-texture ashless paper into a suitable-sized beaker, wash the small precipitate carefully (keeping the funnel covered as much as possible) with an ammonium chloride solution (2%) containing a little freshly prepared ammonium sulphide, and finally suck paper and precipitate dry at the pump. The washed precipitate (*18*) is kept for the identification (seldom for the quantitative determination) of copper, nickel, cobalt, or zinc, by means of spot tests (see Feigl, 1946), whereas the combined filtrate and washings (*19*) are reserved for the determination of the members of the Ammonium oxalate group, calcium and strontium, and of two of the succeeding Ammonium phosphate group, viz. barium and magnesium. To that purpose, boil the combined filtrate and washings (*19*) vigorously until all ammonium sulphide has been expelled, render the solution distinctly acid with HCl (1 + 1), boil again to drive off hydrogen sulphide, filter through a small medium-texture ashless paper to remove most of the free sulphur, wash filter and precipitate with boiling HCl (1 + 1), discard the precipitate (*20*), oxidize the remaining sulphur in the combined filtrate and washings (*21*) by adding a small excess of saturated bromine water, and continue boiling until all trace of bromine is gone (shown by a red colour that persists when adding

2–3 drops of methyl red indicator to the boiling solution) before the separation of the ammonium oxalate group is made as indicated in *c* (below). The elements that may remain in solution are the alkaline earths, magnesium, the alkalies, a small amount of manganese that escaped precipitation as sulphide, and most of the platinum derived from the platinum vessels.

(*b*) If the sample solution contains cobalt, nickel, and zinc in quantities that can be ignored (as usually happen to be present in a portion as small as that taken for the General procedure), and the manganese content does not exceed that of most silicate rocks, a preliminary treatment of the combined ammonium hydroxide filtrate and washings with ammonium sulphide can be omitted. In this case, make the filtrate and washings (*12*) distinctly acid to methyl red with HCl (1 + 1) before the separation of the Ammonium oxalate group is made as detailed in (*c*).

(*c*) Concentrate the combined filtrate and washings (*12*), acidified with hydrochloric acid or else the oxidized and boiled filtrate (*21*) obtained in (*a*), to a volume of about 200 ml, render the clear solution in a 400-ml beaker (marked at the 200 ml level) slightly ammoniacal, heat to boiling, while stirring continuously slowly add 50 ml of a hot solution of ammonium oxalate (4% $(NH_4)_2C_2O_4.H_2O$) and boil for 1–2 min. Cover the beaker with a cover of Pyrex glass, digest near the boiling temperature for 1/2 h, keeping the solution slightly ammoniacal. Allow to cool for 1 h, filter the supernatant liquid through a small properly set close-texture ashless paper (or fritted-glass filter of fine porosity), and wash the precipitate by decantation with 50 ml of a cold dilute ammonium oxalate solution (0.1% $(NH_4)_2C_2O_4.H_2O$) in small portions. (To prevent loss by creeping of the precipitate, consult "Washing of precipitates", p. 43.) Reserve the impure precipitate (*22*) and the filtrate and washings (*23*).

Place the original beaker under the funnel, dissolve the washed impure precipitate in 40 ml HCl (1 + 4), and wash funnel and paper with small portions of hot water. (All of the introduced platinum metals will remain on the paper.) Dilute the filtrate to the mark, add

50 ml of the ammonium oxalate solution (4%, cryst.) and a few drops of methyl red indicator (0.2% alcoholic solution), heat to boiling and, with constant stirring with a rubber-tipped glass rod, add NH_4OH (1 + 1) drop by drop very slowly until the red colour of the solution changes to yellow. Cover the beaker, digest, filter, and wash with dilute ammonium oxalate solution (0.1%, cryst.) as before. Combine the filtrate and washings (*25*) with those already kept for the precipitation of the Ammonium phosphate group (*26*). Reserve the washed precipitate (*24*) for the determination of calcium (and strontium, if any may be present) either *gravimetrically* (*d1, d2*), or *volumetrically* by the permanganate or EDTA methods (*e1, e2*).

(d) Gravimetric method: (1) weighing as oxide

Transfer the paper and precipitate to a weighed platinum crucible with well-fitting cover, dry on the bath, char the paper without inflaming, burn the carbon at a low temperature and under good oxidizing conditions, next cover the crucible and, gradually increasing the temperature, heat until the precipitate is white, Cover the crucible, ignite for 5 min at 1,100–1,200°C, place the well-covered crucible in a desiccator containing sulphuric acid, and weigh as soon as cool. Repeat the weighing after short ignitions until constant weight is obtained. (To prevent possible absorption of moisture during the weighing, keep the weights previously used on the balance pan so that the ignited and coole residue can be quickly reweighed.) If originally present in the sample, besides nearly all of the calcium and a small part of the magnesium as oxides, the weighed residue can be expected to contain the minor part of the manganese if this element is not removed beforehand and all of any strontium, weighed as the oxides Mn_3O_4 and SrO. (Contamination by manganese is indicated by a yellowish to brownish discoloration of the residue; association with strontium is tested with spectroscope after dissolving of the residue in dilute nitric acid.)

Because as a rule, in separating the alkaline earths from magnesium by oxalate, the loss in unprecipitated calcium and the gain in coprecipitated magnesium tend to compensate, giving about correct results for calcium oxide through compensating errors (see p. 137), in ordinary analyses the weight of the ignited residue must be corrected only for any manganese or strontium oxide it may contain if one or both of the oxides should be present in appreciable amounts. In that case, manganese and strontium oxide must be separated from the weighed mixed oxides (*24*) and their amounts therein determined and deducted as follows.

Dissolve the weighed mixed oxides in HNO_3 (1 + 1), transfer the solution to a 25-ml Erlenmeyer flask, evaporate on the steam bath, and remove any manganese present precipitated as oxide by filtration through a fritted-glass filter. (A weighable precipitate is to be reserved, determined and, calculated as Mn_3O_4, deducted from the gross weight of the mixed oxides.) Without spattering evaporate nearly to dryness, and gradually increasing the temperature heat in a radiator for 1 h at 150–160°C. Add to the dry and cool residue 5 ml of absolute alcohol, cork the flask, swirl gently, allow to stand with frequent shaking for 2 h, add 5 ml of absolute ether, cork the flask, swirl gently, and let stand overnight. Next decant the supernatant solution through a suitable small properly set ashless paper into another conical flask, wash the insoluble nitrate several times by decantation with small portions of a mixture of absolute alcohol and absolute ether (1 + 1) until a few drops of the filtrate evaporated on glass leave practically no residue. Dry funnel with paper and contents by gently warming, place the original Erlenmeyer flask containing the main portion of the $Sr(NO_3)_2$ under the funnel, wash funnel and paper dropwise with in total 3 ml of hot water, let drain, and reserve the combined solution and washings.

To these add 2 drops of H_2SO_4 (1 + 1) and 3 ml of absolute alcohol, swirl, let stand 12 h, filter through a small properly set ashless paper into another conical flask, wash paper and precipitate first with dilute alcohol (1 + 1) containing a little sulphuric acid, and then with alcohol (95% v/v) until the free acid is displaced. Transfer paper with contents to a small well-ignited and weighed platinum crucible, dry carefully, char the paper without inflaming, burn the carbon at a low temperature and under good oxidizing conditions, next heat to 1,000°C, let cool in the desiccator, and weigh as $SrSO_4$. (Test spectroscopically for calcium; if present in weighable amount, determine quantity by the sulphate-oxalate method or EDTA method, and make necessary corrections, see p. 145.) Calculate the percentage of calcium oxide, and strontium oxide, of the sample.

(d) Gravimetric method: (2) weighing as sulphate

If desired, calcium may be weighed as the sulphate. The calcium oxalate as starting-point, must be free from strontium and barium (if the amounts of these two elements are not accurately knwon) and the calcium sulphate ignited in such a way that formation of acid sulphate or decomposition by expulsion of SO_3 will be prevented. Therefore ignite the calcium oxalate precipitate (*24*) to oxide as in *d 1* above, cover the crucible, add cautiously under the lid some drops of water by means of a pipet and then a slight excess of H_2SO_4 (1 + 4). Evaporate on the steam bath and next in a radiator to nearby dry, cool, add a few drops of water and evaporate to dryness. Cover the crucible, ignite for 5 min at 1,000°C, cool over sulphuric acid in a desiccator, and weigh as $CaSO_4$. Repeat the weighing after retreatment of the residue with H_2SO_4 (1 + 4), evaporation and short ignition until constant weight is obtained. Calculate the percentage of calcium in the sample in terms of CaO. ($CaO = 0.41193 \times CaSO_4$.)

(e) Volumetric method: (1) oxidimetric by the permanganate method

The use of the permanganate method is limited to precipitates containing calcium oxalate only; strontium, magnesium, and barium oxalates (if their amounts are not accurately known) will be counted as calcium oxalate and must be absent. In that case, carefully wash the calcium oxalate precipitate (*24*) caught on a fritted-glass filter (see p. 141) with cold water to remove adhering ammonium oxalate solution, place filter with precipitate in the original beaker, add 100 ml of water and 25 ml of H_2SO_4 (1 + 4), heat to 70°C or above until decomposition is completem and titrate the liberated oxalic acid at about 60°C with a 0.1 *N* standard $KMnO_4$ solution (p. 102) until a faint pink colour persists without fading for at least 30 sec. Make a blank determination, following the same procedure and using the same amounts of all reagents. Calculate the percentage of calcium (in terms of CaO) of the sample as follows:

$$\text{Calcium oxide \%} = \frac{(A - B) N \times 0.028}{C} \times 100$$

where:

A = ml of standard $KMnO_4$ solution required for titration of the sample;

B = ml of standard $KMnO_4$ solution required for titration of the blank;

N = normality of the standard $KMnO_4$ solution

C = grams of sample used in titration.

(e) Volumetric method: (2) complexometric by the EDTA method

The method is recommended for the determination of very little calcium or magnesium, and for the small sum of both, in presence of large amounts of alkali salts such as in the filtrates (*12,21*) of natural silicates that have been attacked by an alkaline fusion as in the general procedure (Scheme I). The method presupposes absence of interfering elements such as manganese, or zinc, which should be removed beforehand.

Special reagents

(*a*) Sodium hydroxide solution (100 g NaOH per l). – Dissolve 10 g of purest sodium hydroxide pellets in a 200-ml conical flask of resistant glass in 100 ml of recently boiled and cooled water and avoid wetting the top. When all has dissolved, stopper the flask with a one-hole stopper carrying a soda asbestos ("Carbosorb") guard tube, and cool under the tap to room temperature. Transfer the solution to a well-stoppered polyethylene bottle.

(*b*) Ethylenedinitrilo tetraacetate indicator. – Grind together 0.1 g of ethylenedinitrilo tetraacetate and 5 g of anhydrous sodium sulphate in an agate mortar. Keep in dark-coloured bottle.

(*c*) Standard EDTA solution (10.0 g Na_2EDTA per l). – Dissolve 10.0 g of disodium dihydrogen ethylenediamine tetraacetate in 900 ml of water, transfer the solution to a 1-l volumetric flask, dilute with water to the mark, and mix. Standardize against standard calcium solution (reagent *d*), and keep the solution in a well-stoppered polyethylene bottle.

(*d*) Standard calcium solution (1 ml ≡ 0.5 mg CaO). – Dry calcium carbonate, primary standard grade, for 2 h at 280°C, and cool in a desiccator over sulphuric acid. Dissolve 0.9010 g of the dried salt in 50 ml of HCl (1 + 9), transfer the solution to a 1-l volumetric flask, dilute with recently boiled and cooled water to the mark, and mix. (0.5 mg of CaO ≡ 0.35 mg of MgO, see p. 145.)

Procedure

Depending on whether or not preliminary separations are necessary, evaporate the acidified combined filtrate and washings (*12*), or else the oxidized and boiled filtrate (*21*) obtained in *a*, to a volume of about 180 ml, cool to room temperature, transfer the clear solution to a 200-ml volumetric flask, dilute with water to the mark, and mix.

Measure an aliquot (≡ 0.5–5.0 mg CaO) of the filtrate, free from disturbing elements, into a 300-ml conical flask of resistant glass, dilute with water to ca. 50 ml, add 5 ml of sodium hydroxide solution (reagent *a*), swirl, and boil until all ammonia has been expelled. (Test vapour with damped universal indicator paper.) Cool, again add 5 ml of the sodium hydroxide solution (reagent *a*), mix, stopper the flask with a rubber stopper, and let stand ½–1 h. Then add about 30 mg of the pulverised indicator (reagent *b*) and titrate immediately with standard EDTA solution (reagent *c*) until the colour of the mixture changes from pink-red to clear-blue. Take the burette readings and complete the titration if necessary by adding EDTA solution until no further colour change occurs. (The end point tint is stable for only a few min.) Make duplicate determination, and use the average of the readings to calculate the calcium present in the measured aliquots, as CaO, when the EDTA solution is standardized against the standard calcium solution (reagent *d*) which has been carried through the method. Calculate the percentage of CaO in the sample as follows:

$$\text{Calcium oxide } \% = \frac{A \times B}{C} \times 100$$

where:

A = ml of EDTA solution required for titration of the aliquot;

B = normality of the EDTA solution against the standard calcium solution;

C = grams of sample represented in the aliquot used.

(As a rule the same procedure can be followed in preparing the same concentrations on a semimicro-scale.)

B. *After precipitation as phosphate along with magnesium*

(Usually confined to small amounts of calcium in the presence of much magnesium)

If the calcium content is too low to make an appreciable error in the magnesium determination (p. 151) the usual precipitation with ammonium oxalate may be omitted. In that case, after preliminary separations of the members of the Acid, Hydrogen sulphide, Ammonium hydroxide, and Ammonium sulphide groups (if necessary), the small amount of calcium can be precipitated as phosphate along with the magnesium; when present in the filtrates (*12*) or (*21*), acompanied with all of the manganese that has not been removed, and most of any barium and strontium if double precipitations with diammonium phosphate are made.

Where present and heated to 1,100°C, corrections for the small amounts of barium and strontium in the weighed phosphate residue

(*30*) are made on the basis of the normal phosphates, $Ba_3(PO_4)_2$ and $Sr_3(PO_4)_2$, those for manganese on the basis of the pyrophosphate, $Mn_2P_2O_7$. In the rest of the weighed residue (*30*) which can be expected to consist chiefly of $Mg_2P_2O_7$, and of a small quantity of a mixture of $Ca_3(PO_4)_2$ and $Ca_2P_2O_7$ (the amount of the latter being relatively small, may be counted as the normal tricalcium phosphate), calcium can be determined gravimetrically by the sulphate method or volumetrically by the EDTA method as given under Magnesium (p. 144 or p. 157), and its weight, calculated as CaO, added to that of the calcium oxide obtained in *A, d, 1* or *2* (p. 142 or p. 144).

CHAPTER 13

Magnesium

CONSIDERATION OF METHODS

Magnesium comprises over 2.0 per cent of the lithosphere, in which it is combined. Like oxygen, silicon, aluminium, iron, sodium and potassium, magnesium belongs to the major constituents of rocks. It is found as metasilicate (Mg, Ca, etc., silicate) in some pyroxenes and many amphiboles (i.a. in enstatite and hornblende), as orthosilicate in some garnets and chrysolites (i.a. in grossularite and monticellite), as subsilicate in members of the humite group (i.a. in clinohumite and chondrodite), and also in hydrous silicates as mica, clintonite, chlorite, serpentine, talc, and sepiolite.

As a rule magnesium is determined gravimetrically by the pyrophosphate or 8-hydroxyquinoline methods (p. 152 and 155), or volumetrically by back titration or EDTA methods (p. 155 and 157). In almost all silicate rocks the most used method for the separation and determination of appreciable amounts of magnesium, in absence of excessive amounts of phosphorus and ammonium salts, is that as outlined in Scheme I, in which magnesium and most of the barium (the latter only if sulphates were absent and not introduced during the analysis), together with the small amounts of calcium and strontium that may have escaped double precipitation by ammonium oxalate (see pp. 137–141) are separated from the combined ammoniacal filtrates and washings (*23, 25*) by diammonium phosphate. Also the major part of the manganese originally present in the sample and the minor part of the platinum derived from the vessels, can be expected to accompany magnesium if no prior separations of the members of the Hydrogen sulphide and Ammonium sulphide groups have been made (see pp. 50–52, 140).

Instead of gravimetrically by precipitation with diammonium phosphate and weighing as the pyrophosphate, the figure for magnesium may also be determined volumetrically by acidimetric or complexometric methods after dissolving the $MgNH_4PO_4$ precipitate in suitable acids as detailed under *b 2* (p. 155).

Large amounts of magnesium accompanied by small amounts of calcium, strontium, barium, and manganese, are determined more satisfactorily by direct precipitation with diammonium phosphate, omitting the usual prior separation of calcium, strontium and the minor part of manganese, by ammonium oxalate.

Another commonly used method of determination of magnesium in the combined filtrates and washings (*23,25*) from the Ammonium oxalate group, is the 8-hydroxyquinoline method (p. 156) in which magnesium is determined gravimetrically by precipitation in alkaline solution as the quinolate, followed by weighing as the dihydrate, the anhydrous oxyquinolate, or the oxide; or volumetrically by iodometric procedure depending on oxidation of the liberated quinoline by bromate in acid solution.

DETERMINATION OF MAGNESIA IN SILICATE ROCKS IN ABSENCE OF EXCESSIVE AMOUNTS OF PHOSPHORUS, ALKALI SALTS, AND OXALATES

If originally present in the sample and the removal of the members of the Acid, Hydrogen sulphide, Ammonium hydroxide, Ammonium sulphide, and Ammonium oxalate groups, has been properly made, besides foreign alkali salts introduced in the course of the analysis, the two combined filtrates and washings (*23, 25*) from the oxalate precipitation can be expected to contain magnesium and barium together with the small amounts of calcium and strontium that may have escaped precipitation by ammonium oxalate. If the sample contains members of the Ammonium sulphide group which are not removed previously, a small part of the manganese, nickel, and cobalt present, will be caught with the Ammonium oxalate group, whereas their major part present in the combined filtrates and washings (*23, 25*) together with the small amounts of calcium and strontium left in solution, will be precipitated with the magnesium phosphate and must be recovered in analyses of high accuracy. (The actual determinations of the members of the Ammonium sulphide group are better made in separate portions of the sample and their effect deducted from the weighed $Mg_2P_2O_7$ residue.) However, after the removal of all foregoing groups if necessary, the procedure should be varied, depending on the ratio of the magnesium content and the calcium content of the sample. When calcium is preponderant as in most rocks and silicates, magnesium is determined in the combined filtrates and washings (*23, 25*). Double precipitation as phosphate (the last under the most ideal circumstances!) has to be made to obtain a pure magnesium precipitate of definite composition; the second precipitate of magnesium thus obtained, ignited to 1,100°C, and weighed as $Mg_2P_2O_7$, must be tested for its commonest contaminants, viz. calcium and manganese. If present in small determinable amounts, corrections for calcium in the weighed residue may be made on the basis of the normal phosphate, $Ca_3(PO_4)_2$; those for manganese on the basis of the pyrophosphate, $Mn_2P_2O_7$ (see below). Present in larger amounts, the corrections will be too uncertain, and the contaminants must be removed by suitable procedures before the magnesium is precipitated. (Precipitable amounts of barium and strontium are hardly ever present in a portion as small as that taken for the General procedure, and will be left out of account.)

When much magnesium is accompanied by small amounts of calcium, the separation of the Ammonium oxalate group is omitted, and calcium (together with strontium and manganese, when present) is precipitated as phosphate along with the magnesium (and barium, if present) as described under *B* (p. 147). Corrections for the associates in the weighed $Mg_2P_2O_7$ residue have to be made in accordance with the directions given below (p. 153).

After precipitation as phosphate
(Magnesium content <calcium content)

(a) Gravimetric method: weighing as pyrophosphate

Combine the two ammoniacal filtrates and washings (*23, 25*) from the Ammonium oxalate group in a 600-ml beaker of Pyrex glass (marked at the 450 ml level), acidify with HCl (1 + 1), and evaporate the solution to the mark. Cool, add a few drops of methyl red indicator (0.2% alcoholic solution) and 8 g of diammonium phosphate, stir with a glass rod until dissolved, add NH_4OH (1 + 1) very slowly and with constant stirring until the solution is alkaline, stir briskly for 5 min, add while stirring 50 ml of ammonium hydroxide (s.g. 0.90) and, without scraping the sides of the beaker with the glass rod, continue the stirring for 5 min longer. Cover the beaker, and allow to stand in a cool place, preferably overnight. Decant the solution through a properly set 9 cm close-texture ashless paper, wash paper and impure precipitate 5 times by decantation with cool NH_4OH (5 + 95) in small portions, and set the filtrate aside to determine whether any further precipitation takes place; discard the filtrate (*27*) if clear.

Place a 200-ml beaker of Pyrex glass (marked at the 100 ml level) under the funnel, dissolve the impure precipitate (*26*), clinging to the original beaker and stirring rod, in hot HCl (1 + 9), pour the solution through the filter using as little HCl (1 + 9) as possible to obtain complete solution, rinse the original beaker, glass rod, paper and funnel, thoroughly with hot water, discard the paper containing impurities (*28*), cool, and reserve the filtrate (*29*).

Dilute the filtrate (*29*) to the mark, add one drop of methyl red indicator and 0.2 g of diammonium phosphate, stir with a glass rod until dissolved, add NH_4OH (1 + 1) drop by drop very slowly until the magnesium has started to precipirate (critial interval: pH 6.7–6.8), continue the stirring and add NH_4OH (1 + 1) drop by drop until the precipitation is visible complete. Next add 5 ml of ammonium hydroxide (s.g. 0.90), stir vigorously for 5 min, cover the beaker, and allow to stand in a cool place, preferably overnight.

Decant the solution through a fresh properly set 9 cm close-texture ashless paper, wash the precipitate once by decantation with cool NH_4OH (5 + 95), transfer the precipitate quantitatively to the paper, and wash with cool NH_4OH (5 + 95) in small portions until free from chlorides; the precipitate clinging to the beaker should be detached by means of a small piece of ashless paper wrapped round a rubber-tipped glass rod and added to the precipitate. (Complete removal of chloride ions may be tested by collecting a few drops of the last washing in a test tube containing 2 ml of a 0.1 *N* solution of silver nitrate and 1 drop of 4 *N* nitric acid; the white curdy precipitate of AgCl is entirely soluble in ammonia.) Reserve paper and precipitate (*30*) and discard the filtrate (*31*) if clear.

Because with conversion of phosphates to pyrophosphates by heating, phosphates may attack platinum by forming platinum phosphides, transfer the precipitate (*30*) wrapped in its filter to an old weighed platinum crucible with well-fitting cover reserved for that purpose only. Dry on the bath, char the paper slowly without inflaming, burn the carbon at a low temperature and under good oxidizing conditions, next cover the crucible and, gradually increasing the temperature, heat for 15–30 min at 1,000°–1,100°C. Cool the well-covered crucible in a desiccator containing sulphuric acid, and weigh as $Mg_2P_2O_7$; repeat the weighings after short ignitions until constant weight is obtained.

However, when present in the sample and not previously removed, the ignited $Mg_2P_2O_7$ precipitate may contain all of any calcium, chiefly as $Ca_3(PO_4)_2$, which escaped precipitation as oxalate, and all of any manganese, chiefly as $Mn_2P_2O_7$, left in the filtrate from the Ammonium oxalate group. (Precipitable amounts of barium and strontium are rarely encountered in a proportion as small as that taken for the General procedure, and their effects need not be considered in ordinary analyses.) To arrive at a truer figure for MgO, corrections for the usual contaminants in magnesium pyrophosphate should be made as follows.

To correct for calcium, dissolve the ignited precipitate (*30*) in 10 ml H_2SO_4 (1 + 9), transfer the solution to a 150-ml beaker, add

90 ml of ethanol, stir thoroughly for 5 min, and allow to stand overnight in a cool place, while covered. If any precipitate has formed, filter through a 9 cm close-texture paper into a 250-ml flask, wash the paper and precipitate with ethyl alcohol in small portions, and reserve the washed precipitate (*32*) and the combined filtrate and washings (*33*). Transfer the paper and precipitate (*32*) to a small weighed platinum crucible, dry on the bath, char the paper without inflaming, burn the carbon at a low temperature under good oxidizing conditions, next heat to 900°–1,000°C, let cool in the desiccator and weigh as $CaSO_4$, calculate to $Ca_3(PO_4)_2$ using the factor 2.28, and deduct from the weight of the ignited impure phosphate precipitate (*30*).

To correct for manganese, evaporate the alcoholic filtrate (*33*) nearly to dryness, add a few drops of nitric acid, and evaporate to dense white fumes in order to destroy organic matter. Cool, add 70 ml of water, 20 ml of nitric acid, and 10 ml of sulphuric acid in succession, heat the solution almost to boiling, carefully introduce 0.3 g of potassium periodate (KIO_4), boil for 3 min, and then digest just below the boiling point for 15 min to develop the full intensity of colour. Cool, and dilute to a convenient volume. In another 250-ml flask containing the same quantities of reagents treated similarly, match the colour by adding standard potassium permanganate solution, or compare with standard permanganate solution in colorimeter. From the volume standard permanganate solution required, or reading of colorimeter, calculate the amount of $KMnO_4$, in terms of $Mn_2P_2O_7$ using the factor 1.8, deduct from the weight of the ignited impure phosphate precipitate (*30*), and regard, after deduction of both corrections, the difference as $Mg_2P_2O_7$. Calculate the percentage of MgO in the sample as follows:

$$\text{MgO}\,\% = \frac{(\text{A} - 2.28\,\text{B} - 1.8\,\text{C}) \times 0.3623}{\text{D}} \times 100$$

where:

A = weight in grams of impure phosphate precipitate (*30*); B = grams of $CaSO_4$ (*32*); C = grams of $KMnO_4$ (*33*); D = grams of sample used.

After separation of the Ammonium oxalate group

(b) Volumetric method (1) oxidimetric by the 8-hydroxyquinoline method

Principle of method

The use of this method is limited to the combined filtrates (*23, 25*) obtained in the precipitation of calcium as oxalate, in the absence of appreciable amounts of interfering elements, such as copper, zinc, and manganese, which must be removed previously. Magnesium is precipitated as the quinolate, the precipitate dissolved in hydrochloric acid, treated with a measured volume of a standard bromate-bromide solution in slight excess and, after adding potassium iodide, the iodine liberated by the excess of bromine titrated with a standard sodium thiosulphate solution:

$$3\,MgCl_2 + 6\,NH_4OH + 6\,C_9H_7NO \rightarrow 3\,Mg(C_9H_6NO)_2 + 6\,NH_4Cl + 6\,H_2O$$

$$3\,Mg(C_9H_6NO)_2 + 6\,HCl \rightarrow 6\,C_9H_7NO + 3\,MgCl_2$$

$$6\,C_9H_7NO + 24\,HCl + 6\,KBrO_3 + 20\,KBr \rightarrow 6\,C_9H_5NOBr_2 + 2\,KBr + 18\,H_2O + 24\,KCl + 6\,Br_2$$

$$6\,Br_2 + 12\,KI \rightarrow 12\,KBr + 6\,I_2$$

$$6\,I_2 + 12\,Na_2S_2O_3 \rightarrow 12\,NaI + 6\,Na_2S_4O_6$$

Special reagents

(*a*) 8-Hydroxyquinoline solution (25 g C_9H_7NO per l). – Dissolve 2.5 g of 8-hydroxyquinoline in 6 ml of glacial acetic acid, and dilute to 100 ml with water. Prepare fresh as required. Filter if not clear.

(*b*) Standard potassium bromate-bromide solution (0.2 *N*). – Dissolve 2.783 g of $KBrO_3$ (primary standard grade), and 20 g of KBr (free from bromate) in about 300 ml of water in a 500 ml-volumetric flask, dilute to the mark, and mix. Prepare as required.

(*c*) Potassium iodide solution (200 g KI per l). – Dissolve 20 g of potassium iodide in 100 ml of water. Store in the dark.

(*d*) Starch solution (10 g per l).

(*e*) Standard sodium thiosulphate solution (0.2 *N* or less). – Dissolve 24.8 g of $Na_2S_2O_3.5H_2O$ in 500 ml of freshly boiled and cooled water in a sterile dark-coloured bottle, and standardize against the standard $KBrO_3$-KBr solution (reagent *b*) as follows: transfer 300 ml of HCl (1 + 9) to a 500-ml glass-stoppered flask, let stand in ice bath for 20 min, add 25.0 ml of the $KBrO_3$-KBr solution from a pipette, stopper flask, and let stand in ice bath for 10 min longer. Add 20 ml of Kl solution (reagent *c*), swirl, and titrate immediately with the $Na_2S_2O_3$ solution (reagent *e*) from a burette to a straw colour. Add 2 ml of starch solution (reagent *d*), and continue the titration to the disappearance of the blue colour. Calculate the normality of the $Na_2S_2O_3$ solution in terms of the standard $KBrO_3$-KBr solution.

Procedure

Combine the two ammoniacal filtrates and washings (*23,25*) from the Ammonium oxalate group in a 800-ml beaker (marked at the 400-ml level), acidify with HCl (1 + 1), and evaporate the solution to the mark. Cool to 70°C, neutralize the solution with ammonium hydroxide, add in succession while stirring sufficient of the 8-hydroxyquinoline solution (viz. 1 ml of the reagent *a* for each 0.005 g of MgO + 7 ml in excess), and 15 ml of ammonium hydroxide. Stir for 15 min, preferably by an electric stirrer, and let settle. (The supernatant liquid should show a slight yellow colour if enough reagent was added.) Filter, and wash paper and precipitate with hot NH_4OH (2 + 98) in small portions until free from excess of reagent.

Place funnel with contents above the beaker in which the precipitation was made, unfold the paper, place it against the inside of the funnel, and return the precipitate to the beaker by dissolving it from the paper with 200 ml of hot HCl (1 + 4). Cool to room temperature, add 25.0 ml of the standard $KBrO_3$-KBr solution (reagent *b*) from a pipette, stir, let stand for ½ min, add 20 ml of KI solution (reagent *c*), stir, and titrate immediately with the standard $Na_2S_2O_3$ solution, using 2 ml of starch solution (reagent *d*) as an indicator near the end of the titration. (The time from the addition of the KI solution to the end of the titration should not exceed 2 min.) Calculate the percentage of MgO in the sample as follows:

$$\text{Magnesium oxide \%} = \frac{A \times B \times 0.00504}{C} \times 100$$

where:
A = ml of $Na_2S_2O_3$ solution required to titrate the sample;
B = normality of the $Na_2S_2O_3$ solution calculated in terms of the standard $KBrO_3$-KBr solution;
C = grams of sample used in titration.

(Although less satisfactorily, the washed magnesium hydroxyquinoline precipitate obtained in *b 1* (p. 155), can also be collected on a weighed fritted-glass filtering crucible (porosity 4), dried at 130–140°C for 1 h, and weighed as the anhydrous quinolate $Mg(C_6H_6NO)_2$ which contains 12.91 per cent of MgO, or otherwise be ignited under a cover of oxalic acid and weighed as the oxide MgO.)

(b) Volumetric method: (2) complexometric by the EDTA method

In samples containing 0.5–15.0% MgO, the amounts of calcium and magnesium, and the sum of both, in the filtrate from the Ammonium hydroxide group, can mostly be determined by EDTA titration according to the directions given for the determination of these elements in agricultural liming materials (J. Assoc. Offic. Agr. Chemists, 46, 128, 1963).

Special reagents

(*a*) Buffer solution (pH 10). – Dissolve 6.75 g of NH_4Cl in 20 ml of water, add 57 ml of ammonium hydroxide, and dilute with water to 100 ml. Transfer to a well-stoppered polyethylene bottle.

(*b*) Potassium hydroxide-potassium cyanide solution. – Dissolve 28 g of KOH and 6.6 g of KCN in 100 ml of water. Transfer to a well-stoppered polyethylene bottle.

(*c*) Potassium cyanide solution (20 g per l). – Dissolve 2 g of KCN in 100 ml of water. Prepare fresh as required.

(*d*) Eriochrome black T indicator solution. – Dissolve 0.2 g of indicator (Eastman Kodak P6361 or equiv.) in 50 ml of methyl alcohol containing 2 g of hydroxylamine.HCl. Keep in dark-coloured bottle.

(*e*) Standard magnesium solution (1 ml ≡ 0.25 mg Mg). – Dissolve 0.25 g of Mg turnings in HCl (1 + 10), transfer the solution to a 1-l volumetric flask, dilute with double distilled and cooled water to the mark, and mix.

(*f*) Standard calcium solution (1 ml ≡ 1.00 mg Ca). – Dissolve 2.4972 g of calcium carbonate (primary standard grade), previously dried for 2 h at 285°C, in HCl (1 + 10), transfer the solution to a 1-l volumetric flask, dilute with double distilled and cooled water to the mark, and mix.

(*g*) Calcium indicator. – Grind together 1 g of indicator, 10 g of charcoal (Norite A), and 100 g of KCl. (The indicator is described in *Anal. Chem.*, 28 (882), and is available from G.Frederick Smith Chemical Co., Station D, Box 5906, Columbus 22, Ohio.) Keep in dark-coloured bottle.

(*h*) EDTA standard solution (0.4%), for calcium. – Dissolve 4.0 g of disodium dihydrogen ethylenediamine tetraacetate in 900 ml of water, transfer to a 1-l volumetric flask, dilute to the mark with water, and mix. Standardize against the standard calcium solution (see Standardization).

(*i*) EDTA standard solution (0.1%), for magnesium. – Dissolve 1.0 g of disodium dihydrogen ethylenediamine tetraacetate in 900 ml of water, transfer to a 1-l volumetric flask, dilute to the mark with water, and mix. Standardize against the standard magnesium solution (see Standardization).

Standardization

For calcium, pipet 10.0 ml of the standard calcium solution (reagent *f*) into a 300–ml flask and add 100 ml of water. Add 10 ml of KOH-KCN solution (reagent *b*) and 35 mg of Calcein indicator (reagent *g*). Using a magnetic stirrer and artificial light, titrate with the EDTA standard solution (reagent *h*) to disappearance of all green colour. Titrate 3 or more aliquots form the standard calcium solution and use the average to calculate the Ca-titer of the EDTA solution, which = 10/ml EDTA solution.

For Magnesium, pipet 10.0 ml of the standard magnesium solution (reagent *e*) into a 300-ml flask and add 100 ml of water. Add 5 ml of buffer solution (reagent *a*), 2 ml of KCN solution (reagent *c*) and 10 drops of Eriochrome black T indicator. Using a magnetic stirrer and artificial light, titrate with the EDTA standard solution (reagent *i*) until the colour changes permanently from wine red to pure blue. Titrate 3 or more aliquots from the standard magnesium solution and use the average to calculate the Mg-titer of the EDTA solution, which = 10/ml EDTA solution.

Determination

If clear, slightly acidify the combined filtrate and washings (*12*) from the Ammonium hydroxide group with HCl (1 + 1), transfer to a 500-ml volumetric flask, dilute to the mark with water, and mix.

For calcium oxide: pipet a 5–100 ml aliquot (≡ 0.85–1.70 mg MgO) from the adjusted filtrate (*12*) into a 300ml flask, and dilute to about 110 ml with water. Add 10 ml of KOH-KCN solution (reagent *b*) and 35 mg of Calcein indicator (reagent *g*). Using a magnetic stirrer and artificial light, titrate with the 0.4% EDTA standard solution (reagent *h*) to disappearance of all green colour, observing the end point through the solution away from light.

$$\text{Calcium oxide } \% = \frac{A \times B}{C} \times 1.4$$

where:

A = ml of 0.4% Na_2EDTA solution required to titrate the sample;

B = normality of the 0.4% Na_2EDTA solution calculated in terms of the standard calcium solution;
C = grams of sample used in titration.

For magnesium oxide: pipet the same quantity of aliquot (≡ 0.85–1.70 mg MgO) from the adjusted filtrate (*12*) as taken for the CaO-determination into a 300 ml flask, dilute to about 100 ml with water and add the exact quantity of the 0.4% EDTA standard solution (reagent *h*) required for the CaO-determination. Add 5 ml of the buffer solution (reagent *a*), 2 ml of the KCN solution (reagent *c*), and 10 drops of the Eriochrome black T indicator (reagent *d*). Using a magnetic stirrer and artificial light, titrate with the 0.1% EDTA standard solution (reagent *i*) until the colour changes permanently from wine red to pure blue.

$$\text{Magnesium oxide \%} = \frac{A \times B}{C} \times 1.66$$

where:
A = ml of 0.1% Na_2EDTA solution required to titrate the sample;
B = normality of the 0.1% Na_2EDTA solution calculated in terms of the standard magnesium solution;
C = grams of sample used in titration.

CHAPTER 14

Ferrous Iron

CONSIDERATION OF METHODS

The two methods given are suitable for macro and semi-micro analysis respectively. A fusion method for minerals which are insoluble in hydrofluoric acid and therefore required less often, in addition to several other methods, are discussed below with references.

The volumetric method described by Groves (1951, p.88), uses the potassium permanganate titration of ferrous iron brought into solution by the action of hydrofluoric and sulphuric acids. This method assumes no partial oxidation of the ferrous iron during grinding of the sample in air, nor reducing of ferric iron by decomposable sulphides during dissolving in acids. Although the container is usually a platinum crucible, a number of personal variations have been used. The most common modification is that of an inverted polyethylene funnel placed over the crucible while the sample is decomposed on a steam bath; an inert gas is passed into the funnel in order to expel air to obtain an inert atmosphere over the sample (Fig.12).

Fluoride which would interfere in the titration is complexed by the addition of boric acid to the solution before the titration. Variations in the oxidant solution used for the titration have been made; ceric sulphate solution using o-phenanthroline ferrous sulphate complex (ferroin) as indicator has also been found satisfactory (Vincent, 1960).

The spectrophotometric method, using the red coloured complex formed by ferrous, iron with 2,2′-dipyridyl, may be used to advantage for semi-micro analysis (Riley and Williams, 1959a). The method may also be used in macro analysis provided the sample is finely

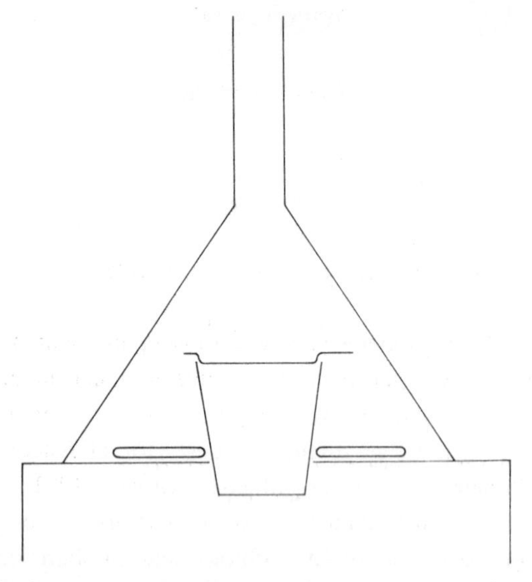

Fig.12. Simple apparatus for digestion of sample under an inert-gas atmosphere.

ground, i.e. <200 mesh and a semi-micro balance is available for weighing out a small portion of the sample, e.g., 5–10 mg. One advantage of this method is that atmospheric oxidation during measurement is avoided since once the reagent has been added, the ferrous iron is complexed and not easily oxidized. A further advantage is that when the absorbance of the solution has been measured, a few crystals of hydroxylamine hydrochloride may be added to the volumetric flask to reduce the ferric iron present in the solution. This allows the total iron to be measured and the value obtained may then be compared with that of the total iron determined elsewhere in the analysis. Using the apparatus shown in Fig.13 (see p. 168) the presence of a completely inert atmosphere allows a prolonged digestion of the sample. This is useful for the solution of minerals which are normally difficult, e.g., garnets.

A method originated by Rowledge (1934) and modified by Groves (1951, p.185) is suitable for a sample which will dissolve only very slowly in hydrofluoric and sulphuric acid, e.g., staurolite, tourmaline and spinel. The sample is fused at an elevated temperature in an inert atmosphere, e.g., N_2, CO_2, using either a gas or electric furnace. The fusion is made with sodium metafluoroborate, the cooled melt is dissolved avoiding oxidation and the ferrous iron titrated either with potassium permanganate or ceric sulphate solution.

The time for solution of the cake in sulphuric acid may be prolonged and the solution may require boiling in an atmosphere of carbon dioxide. A modification which can be used to advantage, since it overcomes these difficulties in solution of the cake, is that given by Hey (1941) in which the cake is dissolved by iodine monochloride in strong hydrochloric acid. Boric acid is not required to be present and the liberated iodine is reoxidized to iodine monochloride using a standard potassium iodate solution as the titrant and carbon tetrachloride as an indicator. This method has also been applied to semi-micro analysis.

The titration method has been modified by dissolving the sample in a polythene bottle containing excess vanadate (Wilson, 1955). The bottle is closed and stored at room temperature, with occasional agitation, for the period required for complete solution of the sample. The excess vanadate is then titrated with ferrous ammonium sulphate solution in the presence of boric acid to complex the fluoride. An advantage of this method is the short manipulating time involved.

In a method described by Wilson (1960), the sample is dissolved in cold hydrofluoric acid containing quinquivalent vanadium. This reagent oxidizes the ferrous iron as it passes into solution. Then, either the excess vanadium is titrated, or the ferrous iron is reformed and determined spectrophotometrically, i.e., as dipyridyl complex. A sample weight of 3–20 mg is taken for a semi-micro determination.

Elimination of interference due to the presence of ca. 4% of carbonaceous matter, has been made possible by the use of the reaction between ferrous iron-bearing solutions and iodine monochloride solution (Heisig, 1928). A large excess of hydrochloric acid is required to be present and the liberated iodine is titrated against standard potassium iodate solution. Use of carbon tetrachloride as an indicator allows the endpoint of the titration to be clearly seen. The detailed procedure has been presented by Nicholls (1960).

For samples soluble in phosphoric acid-pyrophosphate mixture, Ingamells (1960) has described a method of measuring oxygen excess or deficiency. If excess Mn^{2+} is present in the solution, an addition of a standard oxidant is made. The Mn^{3+} remaining in the solution after dissolution of the sample is titrated with ferrous ammonium sulphate solution. Barium diphenylamine sulphate is used as an indicator. The method depends on the stability of Mn^{2+} and Mn^{3+} in the acid mixture.

A recently described method (Van Loon, 1965) uses a pyrex glass flask as a container for the sample, hydrofluoric and sulphuric acids and a known volume of standard potassium iodate solution. The sample is decomposed by heating for 15–20 minutes and the ferrous iron that is released reacts with the iodate to produce iodine. The iodine is volatilized from the boiling solution and the residual iodate after cooling is treated with excess iodide. The liberated iodine is then titrated with standard thiosulphate solution. The amount of thiosulphate and hence the iodate equivalent allows the ferrous iron in the sample to be calculated. In this work experiment showed that no ferrous iron was introduced by the use of a pyrex glass flask for the decomposition. A disadvantage of this method, however, is that attack on the glassware causes a high rate of replacement.

Ferrous iron by the permanganate method

Reagents

Boric acid solution – Dissolve sufficient boric acid in a litre of water to form a saturated solution. Heat the solution to boiling to expel dissolved oxygen, cool and stopper.

Potassium permanganate solution. – Dissolve 1.58 g potassium permanganate in ca. 800 ml of water. Transfer the solution to a 1-l flask and adjust the volume to 1 l with water. Leave the solution to stand for several days if possible before use and then standardize before using by means of the procedure given below.

Procedure

Weigh 0.1–0.5 g of the sample into a 30–40 ml platinum crucible and add 5 ml of hydrofluoric acid. Swirl to mix the sample with the acid. Place 5 ml of water in a separate platinum crucible. To this carefully add 5 ml of sulphuric acid and swirl to mix. Transfer the water and sulphuric acid mixture to the crucible containing the sample and hydrofluoric acid and place a tight fitting lid on the crucible.

Place the crucible immediately over either a low bunsen flame or on a hot plate so that the contents of the crucible simmer for 8–10 minutes. During this period prepare a 600 ml beaker containing 200 ml of cold boiled water, 50 ml of the boric acid solution and 5 ml of sulphuric acid. As soon as the heating period is over, immerse the crucible completely in the solution contained in the beaker. Remove the lid of the immersed crucible with a polythene or glass stirring rod and titrate immediately the ferrous iron with the potassium permanganate solution. The endpoint is reached when a very pale pink colour persists for 30 seconds which indicates a slight excess of permanganate. During the titration, the solution is gently stirred to ensure mixing of the ferrous iron in solution with the titrant.

From the volume and strength of the standardized potassium permanganate solution used in the titration, the weight of ferrous iron in mg present in the sample is calculated.

Standardization of the potassium permanganate

A simple procedure for the standardization of the potassium permanganate (Vogel, 1958, p.274) is as follows: ca. 50 mg of pure iron powder (Specpure) is accurately weighed into a 250 ml conical flask. A rubber bung is placed in the top of the flask. Through this bung is inserted both a glass inlet and an outlet tube. The latter is closed by a bunsen valve. An inert gas, e.g. N_2 or CO_2 is passed through the flask for several minutes. The bung is then loosened and 100 ml of H_2SO_4 (1 + 11) quickly added, the bung being immediately replaced firmly. The reaction is allowed to proceed in the cold. When this is nearing completion the flask is gently warmed until all of the iron powder has dissolved. The flask is then allowed to cool to room temperature with the inert gas supply still connected. When cold, the bung is removed and the sides of the flask washed down with cold boiled water and the solution titrated with the potassium permanganate solution until a very pale pink persists for about 30 seconds. The volume of the permanganate solution equivalent to 1 mg of ferrous iron is then calculated.

Critical considerations

The temperature of the boric acid should be kept as low as possible as this assists in restricting oxidation of the ferrous iron during titration.

If the solution is not stirred during the titration it is possible that local concentrations of ferrous iron may be trapped within the crucible. The addition of ca. 5 ml of phosphoric acid will assist the end-point of the titration.

Interferences

It should be remembered that in this determination it is really the net state of oxidation that is being determined. Carbonaceous matter will interfere giving a high ferrous iron value, while the presence of manganese dioxide will cause a low value to be returned. Graphite does not take part in the reactions and therefore has no effect. The presence of vanadium will cause a positive error, the correction being about twice the value of the vanadium as V_2O_5.

Ferrous iron by the 2,2′-dipyridyl method

Preparation of sample

Solution of the sample is more rapid if the sample is very finely ground. Owing to the oxidation of ferrous iron in the sample if this is done in the dry, acetone is added to the sample in an agate mortar before and during grinding. Usually about 100 mg of sample is ground as finely as possible, dried under an infra red lamp and set aside in a separate small vial for the determination of ferrous iron.

Reagents

2,2′-dipyridyl solution. – Dissolve 0.2 g of 2,2′-dipyridyl in 100 ml of Hcl (1.5 + 98.5).

Sodium acetate solution. – Dissolve 272 g of sodium acetate (hydrated) in 800 ml of warm water and adjust the volume to 1 liter with water.

Acetone solution. – Dissolve 10 ml of acetone to 100 ml with water.

Acid mixture. – Mix 10 ml of hydrofluoric acid with 90 ml of H_2SO_4 (1 + 1) and store in a polythene bottle. Before use, bubble oxygen-free nitrogen through the solution. A useful method of storage for this solution is in a polythene wash bottle with a rubber teat over the outlet when not in use.

If the nitrogen used to obtain an inert atmosphere is not oxygen-free, it should be passed during use through a bubbler containing 15 g of pyrogallol in 100 ml of 50% KOH solution.

Apparatus

A suggested equipment is shown in Fig.13.

Procedure

Weigh 5–10 mg (to the second decimal place) of finely ground

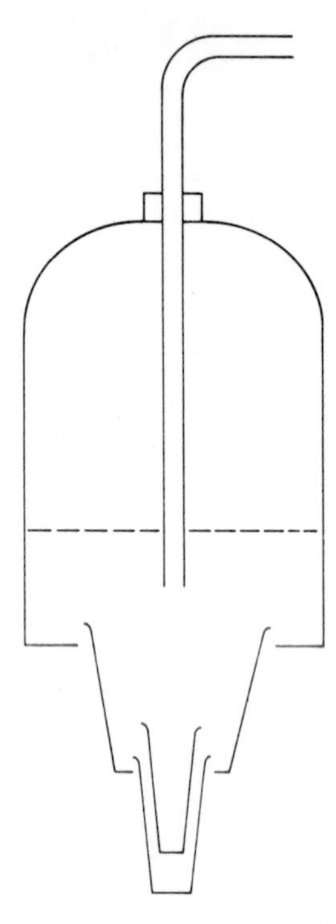

Fig.13. Polyethylene apparatus for digestion of sample for ferrous iron determination.

sample into the bottom polyethylene thimble. Add one or two drops of acetone solution and tap the thimble so that the sample is wetted. Add 3 or 4 drops of acid mixture and lightly insert the top polythene thimble into the bottom thimble containing the sample. This is in order to restrict loss of moisture from the bottom thimble during heating on the water bath.

Place the two thimbles through the hole in the bottom of the cut polythene wash bottle and replace the top portion of the bottle so that it overlaps the bottom by 3–5 mm. Attach the top of the bottle to a clamp, connect the oxygen-free nitrogen supply and allow the gas to pass through the cold assembled apparatus for 5–10 min. The flow of nitrogen should be sufficient to hold a positive pressure inside the apparatus.

After this period of time has elapsed, lower the apparatus over a water bath so that the two polythene thimbles are just inserted into the steam of the water bath. Leave the apparatus in this position for 30–45 min for glasses and normal rocks and for a longer period for samples which contain hydrofluoric acid resistant minerals.

When the sample has been completely decomposed, the apparatus is allowed to cool with the nitrogen still passing. When cold the nitrogen is disconnected and the apparatus taken apart. The outside of the top thimble is carefully washed with cold boiled water directed into a polyethylene funnel which is inserted in a 100-ml

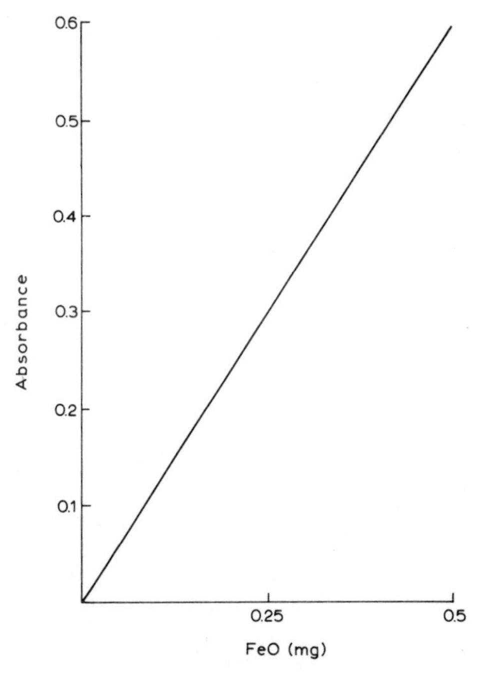

Fig.14. Calibration curve for ferrous iron using a 1-cm cell, 100 ml volume and a wave-length of 522 mμ.

volumetric flask containing 5 ml of 2,2′-dipyridyl solution. The contents of the bottom thimble are then washed into the funnel and the latter rinsed with water. Add 25 ml of sodium acetate solution to the flask and adjust the volume to 100 ml with cold boiled water.

Measure the absorbance of the solution immediately against water in a 1-cm cell using a spectrophotometer with the wavelength set at 522 mμ. Compare the absorbancy of the sample solution with a standard curve. This may be calculated from the readings used for the construction of the total iron curve. The mg of ferric iron should be multiplied by 0.90 to give mg of ferrous iron, (Fig.14).

Critical considerations

When weighing the sample into the polythene thimble, care should be taken that the thimble is only handled by tongs or tweezers otherwise moisture from the hands will cause an error in the weighing.

If the sample has been dried after grinding, due allowance must be made after the H_2O^- has been determined if the result is to be included in a complete analysis made on undried material.

Interferences

These will be the same as for the volumetric titration.

CHAPTER 15

Manganese

CONSIDERATION OF METHODS

Manganese comprises about 0.09% of the lithosphere, in which it is combined. It is found as silicate (rhodonite, pyrosmalite), as oxide (pyrolusite, hausmannite), as hydroxide (psilomelane, manganite), as carbonate (rhodochrosite, manganocalcite), as sulphide (alabandite, hauerite), etc., and is most common in rocks high in iron.

The methods commonly used in the determination of manganese are: (*a*) gravimetric by weighing as the pyrophosphate $Mn_2P_2O_7$ (after prior removal of such elements as would also be precipitated) or as the sulphate $MnSO_4$ (after preliminary separation of the manganese as sulphide from all other substances that yield nonvolatile compounds with sulphuric acid); (*b*) volumetric by titrating a specially prepared solution with an oxidant or reductant; (*c*) colorimetric as permanganate (confined to small amounts of element after preliminary elimination of interfering substances).

Because of the fact that a gravimetric determination of moderate or small amounts of manganese is subject to relative grave errors due to incomplete prior separation or coprecipitation of elements, in silicate rocks the actual determination of total manganese had better be made volumetrically or colorimetrically in a separate portion of the sample, and in the main portion (see Scheme I) only that part, by colorimetry, which is weighed as $Mn_2P_2O_7$ with the magnesium pyrophosphate, as a correction to the magnesia (see p.154). The volumetric methods (bismuthate method, persulphate method) are best for moderate amounts of manganese, whereas in samples in which the amount of manganese present will not exceed 1 per cent, the most widely adopted methods are colorimetric. Based on the violet-red colour given by manganese in a nitric, sulphuric and phosphoric acid solution containing bismuthate, silver nitrate-persulphate or periodate, and having the advantage that (where the manganese content of the sample under test may be unknown or prove to be higher than expected) adjustment of the final volume is possible after development of the soluble coloured complex, the method is almost universally used for the colorimetric determination of small amounts of manganese.

Regarding the oxidants mentioned above: sodium bismuthate is less suitable in colorimetric determinations because the reagent always contains traces of manganese and chlorides, while the insoluble excess causes occlusion, and fading after filtration. The silver nitrate-persulphate oxidation always yields a slight turbidity. Oxidation with periodate (though rather expensive) is to be preferred because the oxidation is free from these objections, no filtration is required, and the violet-red colour is stable for hours, if a slight excess of periodate is used, and reducing agents are absent.

With manganese silicates (like those of the chrysolite group), it is customary to determine the manganese as permanganate in an aliquot of the clear solution of the soluble bases obtained by direct attack of a separate powdered portion of the sample by hydrofluoric and nitric acids (see "Decomposition by acids", p. 35); if not or partly decomposed by acids (like garnet and staurolite), the powdered portion had better be fused with sodium carbonate, the melt transferred to a porcelain dish, dissolved with dilute nitric acid, the solution evaporated to dryness on the steam bath, the residue drenched with warm nitric acid, diluted with hot water, the solution without aid of suction filtered through a fritted-glass crucible of moderate porosity, and the clear filtrate evaporated to dryness on the bath, before proceeding with the determination.

A. Volumetric determination

(For moderate amounts – 10–50 mg – of manganese)

Manganese by the bismuthate method

Principle of method

Oxidation of bivalent manganese by sodium bismuthate in cold nitric acid solution to the heptavalent state and measured reversion to the bivalent state by means of ferrous salt. The reactions are:

$$2\ Mn(NO_3)_2 + 5\ NaBiO_3 + 14\ HNO_3 \rightarrow 2\ NaMnO_4 + 3\ NaNO_3 + 5\ Bi(NO_3)_3 + 7\ H_2O$$

$$2\ NaMnO_4 + 10\ FeSO_4(NH_4)_2SO_4 + 8\ H_2SO_4 \rightarrow 2\ MnSO_4 + Na_2SO_4 + 5\ Fe_2(SO_4)_3 + 10\ (NH_4)_2SO_4 + 8\ H_2O$$

Conditions

Not more than 50 mg of manganese per 100 ml solution should be present; the concentration of nitric acid in the solution should be kept within the limits of 11–22% v/v; at least 28 times as much sodium bismuthate (75%) as manganate must be present.

Interfering substances

Among these may be mentioned: hydrofluoric and nitrous acid, cerium, chromium, cobalt, and vanadium. The first two can be removed by evaporating to dryness and volatilizing, the third is rarely present. The three last-mentioned (if present in amounts over 0.5%) may interfere more or less seriously because chromium is slowly oxidized by permanganate even in the cold, cobalt in nitric acid solution containing iron is oxidized by bismuthate and by permanganate and reduced by ferrous salt, whereas vanadium is reduced by ferrous salt and slowly reoxidized by decinormal permanganate even in the cold. The last three, however, can be removed or caught more or less completely with the Ammonium hydroxide group (see p.51).

Apparatus

Filtering crucible connected with suction flask. A Gooch crucible with asbestos pad or a fritted-glass or porcelain crucible of sufficiently fine porosity to retain all of the bismuthate. Using a Gooch crucible, further purification of the asbestos may be necessary (see Chapter 2). After all soluble iron and chlorine have been washed out, the asbestos should be digested with hot HNO_3 (1 + 2) for 2 h, to make sure of the elimination of the chlorides and to oxidize traces of ferrous iron.

Special reagents

(*a*) Sodium bismuthate. – The sodium bismuthate should contain enough active oxygen to correspond to at least 75% $NaBiO_3$. Manganese and chlorides should not exceed 0.0005 and 0.001%, respectively.

(*b*) Sulphurous acid. – Saturate water with SO_2. Prepare as required.

(*c*) Nitric acid (3 + 97). – Boil 40 ml of nitric acid under a good hood until decolorized, cool, and pass in a current of clean air for 5 min. Mix 30 ml of this acid with 970 ml of water, add 1 g of sodium bismuthate (reagent *a*), shake, and allow to settle. Use the clear supernatant liquid.

(*d*) Ferrous ammonium sulphate solution (ca. 0.1 *N*). – Dissolve 19.6 g of $Fe(NH_4)_2(SO_4)_2.6H_2O$ in 150 ml of cold H_2SO_4 (5 + 95), and dilute to 500 ml with H_2SO_4 (5 + 95). Standardize at the same time the sample is analyzed. Intermediate standard.

(*e*) Standard potassium permanganate solution (0.1 *N* or less). – Prepare and standardize the solution as described under Total Iron, pp.102–103 (1 ml 0.1 *N* $KMnO_4 \equiv$ 1.10 mg Mn $\equiv$ 1.42 mg MnO).

Procedure

When the amount of manganese is too small for satisfactory titration, and no suitable adjustments can be made in order to produce the desired concentration, the percentage of manganese in a rock or mineral should be determined by the use of more refined techniques as e.g., spectrophotmetric methods (Chapter 3).

The titrimetric determination of manganese in the sample should be carried out as follows.

After evaporation of the clear filtrate to dryness (see p. 172), dissolve the residue in 50 ml of hot HNO_3 (1 + 3), transfer the solution (or a measured cooled aliquot diluted to 50 ml with the same acid) that will contain 10–50 mg of manganese, to a 300-ml Erlenmeyer flask. Cool, add about 0.5 g of sodium bismuthate (reagent *a*), and boil for 2–3 min. Add sufficient sulphurous acid (reagent *b*) to destroy the permanganate colour and excess of sodium bismuthate, and continue the boiling for several minutes longer. Cool the solution to 10–15°C, add 0.5 g of sodium bismuthate (or an amount equal to at least 28 times the weight of manganese present), and agitate the mixture vigorously for 1 min. Add 50 ml of cold dilute nitric acid (reagent *c*) and filter the solution, with the aid of suction, through a Gooch crucible or fritted-glass or porcelain crucible of sufficiently fine porosity. Wash with cold dilute nitric acid (reagent *c*) until the pink colour has been discharged, plus 2–3 ml in excess. Titrate the excess of ferrous sulphate with standard potassium permanganate solution (reagent *e*) to the appearance of a faint pink colour.

Blank. Make a blank determination, following the same procedure and using the same amounts of all reagents and washings, including the exact volume of the ferrous ammonium sulphate solution that was used with the sample, and titrate with the standard potassium permangante solution.

Calculate the percentage of manganese (or of manganous oxide) of the sample as follows:

$$\text{Manganese } \% = \frac{(A - B)C \times 0.0110}{D} \times 100$$

$$\text{Manganous oxide } \% = \frac{(A - B)C \times 0.0142}{D} \times 100$$

where:

A = ml of $KMnO_4$ solution required to titrate the blank;

B = ml of $KMnO_4$ solution required to titrate the sample;
C = normality of the $KMnO_4$ solution;
D = grams of sample used.

B. Spectrophotometric determination
(For small amounts –0.05–0.5 mg – of manganese)

Manganese by the periodate method

Principle of method

Oxidation of manganese in acid solution by means of potassium periodate. Photometric measurement is made at approximately 525 mμ.

Concentration range

The recommended concentration range is from 0.05 to 0.5 mg of manganese in 100 ml of solution, using a cell-depth of 2 cm.

Stability of colour

The permanganate colour is stable for hours if reducing agents are absent.

Interfering elements

Fe, Cr, and Ti, less than 2.0 mg, 1.0 mg, and 0.5 mg, respectively, per 100 ml of solution, do not interfere. Amounts exceeding these maximum limits can be removed or caught more or less completely with the Ammonium hydroxide group.

Special reagents

(*a*) Sodium sulphite solution (100 g Na_2SO_3 per l). – Dissolve 5 g of pure anhydrous sodium sulphite in 50 ml of water, and filter through fritted-glass filter if cloudy. Prepare fresh each time.

(*b*) Standard manganese solution (1 ml ≡ 0.05 mg Mn ≡ 0.065 mg MnO). – Dissolve 0.7200 g of purest potassium permanganate in

100 ml of water, add 10 ml of H_2SO_4 (1 + 1), and reduce the permanganate by addition of sodium sulphite solution (reagent *a*). Boil to remove excess of SO_2, cool, dilute in 500-ml volumetric flask to the mark with water, and mix. Add 50 ml of this solution to a 500-ml volumetric flask, dilute to the mark with water, and mix. Keep away from light, protect from dust and other reducing substances.

(*c*) Potassium periodate solution (10 g per l). – Dissolve 2.5 g of potassium periodate in a 300-ml beaker (marked at the 250 ml level) in ca. 240 ml of hot water, cool to room temperature, add 2.5 ml of HNO_3 (1 + 1), dilute to the mark with water, and mix.

(*d*) Sodium nitrite solution (20 g $NaNO_2$ per l). – Dissolve 0.2 g of sodium nitrite in 10 ml of water. Prepare fresh each time.

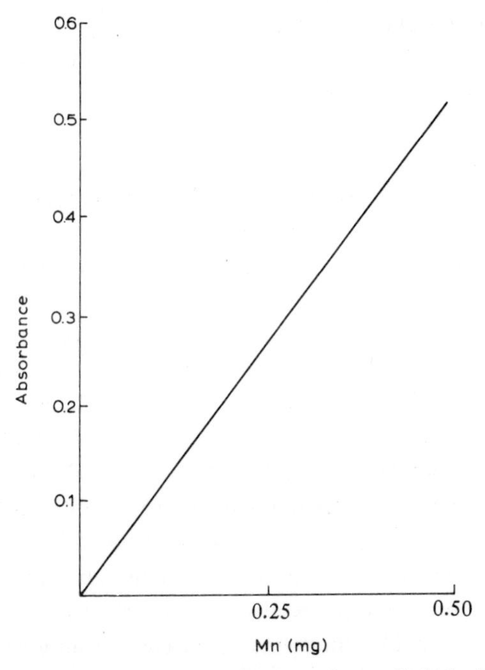

Fig.15. Calibration curve for manganese using a 2-cm cell, 100 ml volume and a wave-length of 525 mμ.

Preparation of calibration curve

Transfer 1.0, 2.0, 5.0, 7.0, 9.0, and 10.0 ml of the manganese solution (reagent *b*) to six 150-ml beakers (marked at the 10-ml level), adjust to the mark by addition of water or by evaporation, and cool. To each beaker, and to an additional beaker for the reference solution containing 10 ml of water, add 10 ml of nitric acid, 5 ml of phosphoric acid, and 10 ml of potassium periodate solution (reagent *c*). Heat to boiling, boil gently for 2 min, digest just below the boiling point on the steam bath for 10 min to develop full intensity of the colour. Cool, transfer to 100-ml volumetric flasks, dilute to the mark with water, and mix.

Transfer a suitable portion of the reference solution to an absorption cell having a 2-cm light path, and adjust the photometer to the initial setting, using a light band centered at approximately 525 μ. Maintaining this photometer adjustment, take the photometric readings of the calibration solutions. Plot the values obtained against mg of Mn per 100 ml of solution (Fig.15).

Procedure

Following this procedure, also make a blank determination (B), using the same amounts of all reagents.

After evaporation of the clear filtrate to dryness (see p.172), dissolve the salts in 50 ml of hot HNO_3 (1 + 3), cool, transfer to a 100-ml volumetric flask, dilute to the mark with water, and mix.

Transfer two like amounts of aliquot ($\equiv$ 0.05–0.5 mg) of the solution obtained, to two 150-ml beakers. To each beaker add 10 ml of nitric acid, 5 ml of phosphoric acid, and 10 ml of potassium periodate solution (reagent *c*). Heat to boiling, boil gently for 2 min, digest just below the boiling point on the steam bath for 10 min. Cool, transfer to two 100-ml volumetric flasks, dilute to the mark with water, and mix; use one of the diluted aliquots for use as a reference solution (R), and reserve the other one (S).

Transfer a suitable amount of the reference solution (R) to a dry 150-ml beaker, while stirring add sodium nitrite solution (reagent *d*) dropwise until the permanganate is completely reduced. Next transfer a suitable portion of the treated reference solution to the absorption cell used before, and take the photometric reading of the solution (S) as described in "Preparation of calibration curve".

Convert the photometric readings of the final sample and blank solutions to mg of manganese by means of the calibration curve. Calculate the percentage of manganese (or of manganous oxide) as follows:

$$\text{Manganese } \% = \frac{A - B}{C \times 10}$$

$$\text{Manganous oxide } \% = \frac{(A - B) \times 1.29}{C \times 10}$$

where:

A = mg of manganese found in the aliquot used;
B = reagent blank correction in mg of manganese;
C = grams of sample represented in the aliquot used.

CHAPTER 16

Chromium

CONSIDERATION OF METHODS

Chromium is widely diffused in nature. It occurs as oxide (chromite, picotite) as an accessory constituent in many basic igneous rocks, especially in peridotite rocks and the serpentinites derived from them as the result of the crystallization of a magma, very low in silica, high in magnesia and containing alumina, where the magnesium may be in part replaced by ferrous iron and manganese, the alumina by ferric iron and chromium, and vice versa; as silicate (uvarovite); as chromate (crocoite); as hydrous sulphate (knoxvillite); as iron-chromium sulphide (daubréelite) in some meteoric irons, etc.

Because of the fact that a gravimetric determination of moderate or small amounts of chromium is subject to relative grave errors due to, e.g., incomplete prior separation or coprecipitation of elements, in general the actual determination of total chromium in rocks had better be made volumetrically or colorimetrically in a separate portion of the sample, and in the main portion (see Scheme I, p. 85) also that part, by colorimetry, which is weighed as Cr_2O_3 as a correction to the alumina.

Most widely adopted methods for the determination of chromium are based on its preliminary oxidation to the chromate. Moderate amounts of it are generally determined volumetrically by some method in which the chromate is, e.g., reduced to the trivalent state by ferrous sulphate followed by measured oxidation with permanganate or is determined by direct potentiometric titration with ferrous sulphate; small amounts (less than 1 mg) are usually determined colorimetrically by the chromate or the diphenylcarbazide method.

The oxidation of chromium from the trivalent to the sexavalent state can be made by dry or wet attack. The former can be done by fusion with alkali carbonate in an oxidizing atmosphere or with sodium peroxide, whereas the last can be accomplished: (*a*) in alkaline solution by potassium persulphate, sodium or hydrogen peroxide, bromine, permanganate, percarbonate or perborate; (*b*) in acid solution by the use of persulphate with catalysis by silver nitrate, of lead dioxide, nitric, bromic, chloric, perchloric or permanganic acid. The reduction of chromates and dichromates to chromic salts may be effected by numerous reagents, e.g., by ferrous salts, hydrogen peroxide in acid solution, alcohol, sulphides, and sulphites.

The outstanding method of separating chromium in solution from iron, manganese, nickel, cobalt, and elements such as titanium or zirconium, is based on the precipitation with sodium hydroxide with the aid of sodium peroxide; the same result can be obtained by fusing with sodium peroxide and extraction of the melt with water. These treatments leave chromium in the filtrate still associated with elements such as silicon, aluminium, vanadium, uranium, molybdenum, tungsten, and phosphorus. All of these foreign elements, save a part of the silicon, can be removed by reducing the chromium to the trivalent state and reprecipitating with sodium hydroxide and carbonate. However, complete separation of chromium from all other elements present in rocks and ores containing not more than only a few per cent, is not required if volumetric or colorimetric procedures are applied. Because of this, and also the fact that in rocks chromium is mostly present in only small amounts ($<0.5\%$), this element is usually determined colorimetrically in a separate portion of the sample by quick procedures that are subject to interference by but few other elements.

Amounts, ranging from 0.5–5 mg Cr per 100 ml of solution, can be matched visually by the chromate method in Nessler tubes or more accurate by means of colorimeters of the Duboscq or Pulfrich type equipped with suitable light filters. Microgram quantities, within the range of 0.005–0.1 mg Cr per 100 ml, are determined spectrophotometrically by the diphenylcarbazide method using a cell depth of 2 cm. (Cells having other dimensions may also be used, provided suitable adjustments can be made in the amounts of sample and reagents used.)

CHROME MINERALS

Because iron and nickel crucibles are strongly attacked by sodium peroxide, and zirconium crucibles (although more resistant) may contain small amounts of chromium, in this procedure porcelain crucibles should be used.

Transfer about 5 g of dry yellow sodium peroxide (low in chromium) to a 30-ml thick-walled porcelain crucible with cover. Add 0.5 g of the finely ground air-dry sample, mix well using a platinum rod, carefully clean the rod of adhering particles by scraping with another rod, cover the mixture with a layer of about 2 g of sodium peroxide, dry, and fuse as described under Decomposition by fluxes (pp. 32–34). After fusion, allow the melt to cool, transfer the whole to a 500-ml Pyrex beaker with cover glass, while covered add about 200 ml of water, tilt the crucible, and keep the beaker covered until the action has moderated. Next warm until the melt has dissolved, wash any particles adhering to the crucible into the beaker, remove crucible and cover, filter through No.54 Whatman paper (previously thoroughly washed with a 5% solution of NaOH to remove soluble organic matter), wash the residue with hot water, and reserve the filtrate and washings.

Ignite the residue in the same porcelain crucible, recover any chromium in the ash by a second fusion with sodium peroxide and filtration, and combine the two alkaline filtrates and washings. Add 1 g of sodium peroxide, allow to stand on the steam bath for ½ h, filter through an asbestos pad (preferably on a small Buchner funnel), and wash beaker and filterpad with a cold 2% solution of NaOH. Combine the filtrates and washings, and determine chromium colorimetrically according the chromate method (*A*).

Chromite. Fuse with pyrosulphate in silica crucible, dissolve the melt in dilute acid, filter, fuse any unattacked material with anhydrous sodium carbonate in platinum crucible, dissolve the melt in dilute sulphuric acid, filter, and combine the filtrates. Remove interfering elements by effective separations before the determination of chromium is attempted.

A. Chromate method

(For small amounts – 0.5–5 mg – of chromium)

Principle of method

Oxidation of chromium in alkaline solution to the sexavalent state. The yellow colour of sample solution is compared with colour standards.

Concentration range

0.5–5 mg of chromium per 100 ml of solution.

Stability of colour

The colour is stable for at least 1 h.

Interfering elements

Appreciable amounts of U, Cu, and Pt, which elements form soluble coloured compounds in 5% solution of NaOH containing peroxide, may interfere.

Apparatus

Measurement can be made visually in Nessler tubes or by means of colorimeters of the Duboscq or Pulfrich type equipped with suitable light filters.

Special reagents

(*a*) Standard potassium dichromate solution (1 ml ≡ 0.1 mg Cr ≡ 0.146 mg Cr_2O_3). – Dissolve 0.283 g of $K_2Cr_2O_7$ (primary standard grade) in water, dilute to 1 liter in a volumetric flask, and mix.

(*b*) Colour standards. – Prepare solutions containing approximately the same concentration of chromium and alkali as the sample to be compared with them. The colour standards should be freshly prepared.

Procedure

Compare the yellow colour of the final filtrate, containing 0.5–5 mg per 100 ml of solution, with the colour standards (reagents *b*).

ROCKS

In the usual case, fuse 1–5 g of the finely ground air-dry sample in a platinum crucible with the fourfold quantity of anhydrous sodium carbonate and a little potassium nitrate. (The latter to an amount little greater than needed to oxidize fully the oxidizable components of the sample to prevent the crucible being attacked.)

Extract the melt with water, add a few drops of ethyl alcohol to reduce the sodium permanganate, filter through No.54 Whatman paper (previously thoroughly washed with a 5% solution of NaOH to remove soluble organic matter), carefully wash paper and residue with a 1% solution of Na_2CO_3, and reserve the filtrate and washings. When fusion was incomplete, repeat the operation, and combine the alkaline filtrates and washings. (In case the colour of the combined filtrate and washings is to faint to match, concentrate as much as possible by evaporation in a porcelain dish on the steam bath.)

When present in quantities of 0.5–5 mg Cr per 100 ml of solution, determine chromium in accordance with method *A*; determine lower concentraties of chromium photometrically as described in method *B*.

B. Diphenylcarbazide method
(For very small amounts – 0.005–0.1 mg – of chromium)

Principle of method

Chromium in the sexavalent state, forms a soluble red-violet complex with diphenylcarbazide. Photometric measurement is made at approximately 540 mμ.

Concentration range

The recommended concentration range is from 0.005 to 0.1 mg of chromium in 100 ml of solution, using a cell depth of 2 cm.

Stability of colour

The colour of the chromium complex develops immediately but starts to fade within a short period of time. Photometric measurement should be made within 5 min after addition of the diphenylcarbazide.

Interfering elements

Appreciable amounts of Cu, V, and Mo, which may have escaped precipitation in a 5% solution of NaOH, forming soluble coloured compounds, will interfere. Oxidizing agents cause rapid fading of the complex.

Special reagents

(*a*) Standard potassium dichromate solution (1 ml ≡ 0.1 mg Cr ≡

0.146 mg Cr_2O_3). – Dissolve 0.283 g of $K_2Cr_2O_7$ (primary standard grade) in water, dilute to 1 liter in a volumetric flask, and mix.

(*b*) Ammonium persulphate $(NH_4)_2S_2O_8$.

(*c*) Silver nitrate solution 1%. – Dissolve 1.09 g $AgNO_3$ in ca. 80 ml of water, dilute to 100 ml and mix.

(*d*) Diphenylcarbazide solution (10 g per l). – Dissolve 0.20 g of $[CO(NH.NH.C_6H_5)_2]$ in 20 ml of methanol. The solution decomposes within a few hours and should be freshly prepared.

Preparation of calibration curve

Dilute solution (*a*) given above, tenfold (1 ml ≡ 0.01 mg Cr) and transfer 1.0, 2.0, 5.0, 7.0 and 10.0 ml to five 200-ml Erlenmeyer

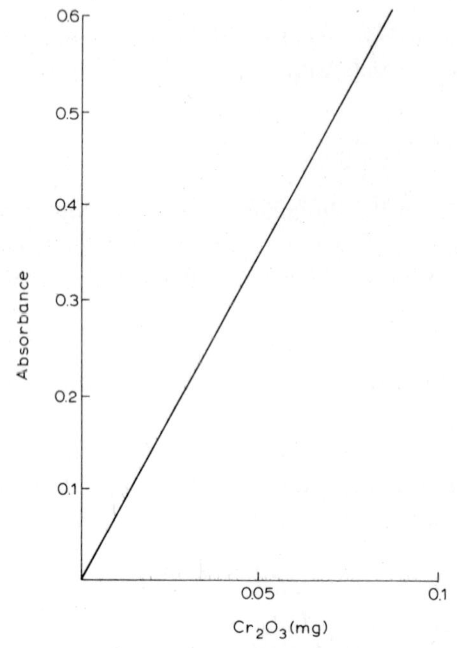

Fig.16. Calubration curve for chromium using a 2-cm cell, 100 ml volume and a wave-lenght of 540 mμ.

flasks. To each flask and to an additonal flask for the reference solution, add 20 ml of H_2SO_4 (1 + 98), 0.5 g of ammonium persulphate (reagent *b*) and two drops of silver nitrate solution (reagent *c*). Heat gently to boiling and continue boiling for 10 minutes in order to oxidize the chromium to chromate and to destroy any excess of persulphate and peroxide formed. Cool to room temperature, transfer to 100-ml volumetric flasks, dilute to the mark with water and mix.

Transfer a 5-ml aliquot of each solution to a 100-ml volumetric flask, dilute to about 90 ml with water, mix, add 1.0 ml of diphenylcarbazide solution (reagent *c*), dilute to the mark with water, and mix.

Finally transfer a suitable portion of the reference solution to an absorption cell having a 2-cm light path, and adjust the photometer to the initial setting, using a light band centered at approximately 540 mμ. Maintaining this photometer adjustment, take the photometric readings of the calibration solutions. Plot the values obtained against mg of Cr per 100 ml of solution (=5% of the original amount added, Fig.16).

Procedure

Following the same procedure, make a blank determination using the same amounts of all reagents, for use as a reference solution.

Evaporate the combined filtrate and washings as obtained above (p. 184) to a volume of about 200 ml in a porcelain dish on the steam bath. Cool to room temperature, filter the solution through paper into a 250-ml volumetric flask, wash dish and paper with a 1% solution of Na_2CO_3, dilute the solution to the mark with water, and mix.

Transfer a 25-ml aliquot of the solution to a 200-ml conical flask, add 2 drops of methyl orange (0.1% aqueous solution), neutralize with H_2SO_4 (1 + 1), add 20 ml of H_2SO_4 (1 + 98), 0.5 g of ammonium persulphate (reagent *b*) and two drops of silver nitrate solution (reagent *c*). Heat gently to boiling and continue boiling ten minutes in order to oxidize the chromium and to destroy any excess of persulphate and peroxide formed. When manganese is present, full development of the permanganate colour indicates that oxidation is complete. Allow the solution to cool and then neutralize it by small additions of solid sodium carbonate, Centrifuge off the precipitate which consists of iron and any silver chloride that has formed, alternatively remove the precipitate by filtering through a filter paper previously washed with dilute sodium carbonate solution (1%). Acidify the filtrate with H_2SO_4 (1 + 6), add 20 ml H_2SO_4 (1 + 98) and transfer to a 100-ml volumetric flask, dilute to the mark with water and mix.

Transfer a 5-ml aliquot of this solution to a 100-ml volumetric flask, dilute to about 90 ml with water, mix, add 1.0 ml of diphenylcarbazide solution (reagent *c*), dilute to the mark with water, and mix.

Immediately transfer a suitable portion of the reference solution to the absorption cell used before, and take the photometric reading of the sample solution, as described in "Preparation of calibration curve".

Convert the photometric reading of the final solution to mg of chromium (or of chromium oxide) as follows:

$$\text{Chromium }\% = \frac{A}{B \times 10}$$

$$\text{Chromium oxide }\% = \frac{A \times 1.46}{B \times 10}$$

where:

A = mg of chromium found in 100 ml of the final solution;

B = grams of sample represented in 100 ml of the final solution.

AMMONIA PRECIPITATES

Recover any chromium in the ignited and weighed ammonia precipitates by fusion with anhydrous sodium carbonate and a little nitrate, extracting the melt with water, adding a few drops of ethyl alcohol, and filtering, as described for rocks. In the combined filtrate and washings, determine chromium colorimetrically in accordance with the method A or B.

OTHER METHODS

Chromium forms a stable violet coloured complex with a DCTA (1,2-diaminocyclohexane tetraacetic acid) solution when boiled (Selmer-Olsen, 1962). Certain other metals e.g., Cu, Mn, Fe, and Co, also produce coloured complexes and their presence requires the use of special procedures.

Spectrophotometric determination of chromium, using the violet chromium-ethylenedinitrilotetraacetate complex, has been described for quantities of 1–15 mg Cr_2O_3 in the aliquot (Kameswara et al., 1965). In this work, sodium bicarbonate is used to adjust the pH and this reagent also behaves as a catalyst for the formation of the complex. The separation of other members of the Ammonium group which may interfere, has been given in a previous paper (Den Boef et al., 1960) together with a detailed account of interferences.

CHAPTER 17

Alkalies

CONSIDERATION OF METHODS

The flame photometric determination of the alkalies has replaced the classical method of Lawrence Smith because of the advantage of reliability and speed (Washington, 1930). Quantities as small as 0.01 per cent of both sodium and potassium can be determined with smaller errors than would be possible by a gravimetric procedure. Determination of these two constituents, after decomposition of the sample by treating with hydrofluoric and sulphuric acids, is both rapid and convenient since minor constituents like Ti, Mn, Cu, Ni, Co, Zn, V, Cr, Zr, Ba, Sr, and Li, may also be determined on the same solution.

Flamephotometry utilizes the excitation of these elements when they are sprayed in solution form into a flame. A detailed study of the physical processes involved in this technique has been given by Dean, 1960, p.7). Measurement of the emission intensity is made either by a direct reading galvanometer or by photocells and a photomultiplier.

The simplest instruments (e.g., Eel Flamephotometer) use light filters to select the region of the spectrum in which suitable spectral lines occur for the elements concerned and to prohibit radiation emitted by the other constituents of the sample solution. Monochromators are also used (e.g., Unicam 900, Beckman D U Flamephotometer attachment) which allow a finer selection of the spectral lines and therefore enable measurement of the emission intensity on either side of the spectral line.

Each instrument has its own particular design of burner and spray injection system; different burners being used for specific gas mixtures. Commonly used gas mixtures include: oxygen–hydrogen, air–acetylene, air–coal gas and air–propane. The lower temperature flames excite fewer elements and therefore are generally used in the determination of sodium and potassium, these being excited at a lower temperature.

The problem of interference by other constituents in the sample solution has been dealt with by various procedures which are given below:

a) Preliminary group separation.

b) Addition of similar concentrations of interfering elements to the standard.

c) Radiation buffers.

d) Internal standards.

e) Standard additions.

f) Ion exchange resin separation of interfering elements.

g) Scanning of emission peak and deduction of background.

The two main causes of interference in the determination of sodium and potassium are:

(*1*) Emission by other constituents which enhances the emission of the constituent to be measured. Examples of interference by enhancement are the two commonly occuring elements iron and calcium. These produce an intense emission of their own over a wide portion of the spectrum near the peak emission of sodium and potassium.

(*2*) Depression is caused by aluminium and magnesium if these are present in sufficient quantity.

Preliminary group separation

Both iron and aluminium may be removed by precipitating the Ammonium hydroxide group (Brannock and Berthold, 1949), using gaseous ammonia to avoid the addition of contaminants. The suppression of calcium interference in the determination of sodium using a filter type instrument has been effected by the addition of aluminium to the sample solution (Williams, 1960). If it is required to remove the calcium, the separation may be made by the addition of a suitable quantity of solid oxalic acid to the sample solution before the passage of the ammonia used to precipitate the Ammonium hydroxide group. Both precipitates are then centrifuged off together and discarded.

Addition of similar concentrations of interfering elements to the standards

The principle used is that similar enhancement or depression of the emission will be obtained by having a similar composition for the standards as in the sample solution. The addition of the same concentrations of the interfering constituents to the standards requires a detailed knowledge of the major constituents in the sample. The standard of purity required in the reagents which are added must be high in order to avoid adding unknown quantities of sodium and potassium to the standard solutions. The solutions added should contain the same anions as those present in the sample solution. If complete analyses are required of a number of samples, it is convenient to leave the determination of sodium and potassium until after the major constituents are known.

Standard concentrations of the interfering constituents are prepared at a high concentration, e.g., 10 mg per ml, stored in polythene bottles and added in the necessary concentration to the standard sodium and potassium solutions. This procedure tends to be more time consuming than the standard addition procedure.

Radiation buffers

The radiation buffer procedure uses the principle that interference is cancelled out by addition of high concentrations of the interfering constituents in both sample and standards. The interference reaches a constant maximum if sufficient of the constituent is present in the solution. A radiation buffer containing Al, Fe, Ca and Mg for use with all types of silicate material has been described by Penner and Inman (1961).

This procedure however, is not satisfactory when only small quantities of sodium and potassium are present since the total emission shown by the sample is large in relation to that due to the desired constituent.

Internal standards

Another method of combating the effects of interference is by use of an internal standard. The principle here is that any effects on the constituent to be measured will be similar to that on the internal standard, e.g., lithium. Emission of the sample solution is therefore measured as a ratio of the intensity of the emission exhibited by the internal standard. A lithium internal standard (200 p.p.m.) has been suggested for use on a Perkin Elmer 52A in the determination of potassium (Cooper, 1963).

Standard addition

The standard addition technique has been widely used owing to its simplicity, particularly when the bulk composition of the sample is unknown. The emission of a sample solution is measured and then a sample solution is measured to which a known quantity of the same constituent has been added. The calculation is made using the principle that the emission of the standard added to the sample solution will be effected in a similar manner and extent to that already present in the sample solution. Either a single addition is made or a series of additions of different strengths, the latter being more accurate. When using this method it is essential that the background emission contributes only a negligible portion of the total emission, otherwise it becomes necessary for the peaks to be scanned and the background emission subtracted.

Ion exchange separation of interfering elements

An ion exchange resin separation of interfering constituents has been described (Riley, 1958a) in which iron, aluminium and titanium were retained on an Amberlite IRA–400 (Cl) analytical grade, citrate form. The sodium and potassium passed through the column and were collected for measurement. In this work an Eel flame-photometer was used. After the volume of the solution had been adjusted to a known volume, sodium was measured in the first instance and the necessary correction for the sodium light passing through the potassium filter calculated ready for deduction from the subsequent potassium measurement.

Scanning of emission peak and deduction of background

When measuring small quantities, e.g., 0.1–1.0 p.p.m. in the sample solution, it is necessary for the instrument to be such that the spectral peak may be scanned in order to measure and subtract the emission due to the other constituents in the sample solution before the calculations are made (Fig.17). Examination of the spectral peak and subtraction of the background has been succesfully used on a Beckman D U instrument using an oxygen–hydrogen flame (Easton and Lovering, 1964). Measurements were made on either side of the respective peaks and since the sodium peak is symmetrical the background was measured at 580 or 600 mμ. The potassium peak however, was not found to be symmetrical at low concentrations requiring a wide slit width and was measured at 725 and 825 mμ. The background at 768 mμ was found graphically.

Where low concentrations are measured, it is inevitable that the slit width (controlling the quantity of light falling upon the photo-

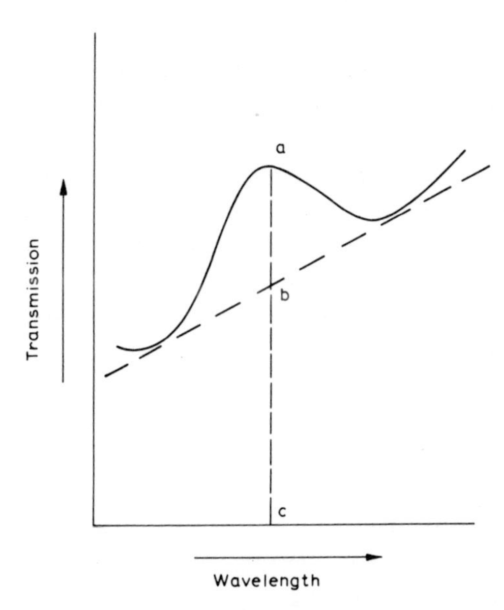

Fig.17. Transmission curve of a sample solution. The background emission is measured by taking readings either side of the peak (*a*). The transmission (*bc*) is deducted from *ac* to obtain the true emission of the solution.

cell or photomultiplier) must be greatly increased thus increasing the background emission. For this reason standards, although free from interfering constituents, must also be examined by taking measurements either side of the spectral peak.

Physical factors

The physical factors affecting the emission include viscosity, surface tension and temperature; the latter exerting influence upon the former two properties of the sample solution. An increase in viscosity of the solution will decrease the emission while a higher temperature will decrease the viscosity of the solution, thereby increasing the emission. Where a spray chamber forms part of the injection system, a lower surface tension will allow the formation of smaller spray particles less likely to settle out on the walls of the vessel. This will increase the emission exhibited by the solution.

Presence of certain anions may also influence the spray particle size. Sulphuric, hydrochloric and perchloric acids each have an effect when present in the sample solution (Cooper, 1963). The decrease in emission was shown to depend upon the concentration of the acid, at least at low concentrations, e.g. $<3\%$.

Salts such as ammonium sulphate and carbonate will also affect viscosity and surface tension, the magnitude of the effect depending upon their concentration. For the reasons given above it is necessary to prepare the standards, against which the sample solution is to be measured; in such a manner that the concentration of acid and salts is similar in concentration.

It is most important when considering conclusions in the literature on the effect of interferences and the various procedures to eliminate them, to realise that such factors as the instrument design, flame composition and the construction of the burner, will all influence the results from which the conclusions have been drawn.

Preparation of the sample solution

Weigh up to 1 g of the air-dry sample (–200 mesh to the linear inch) in an open, well cleaned, well ignited, and cool platinum crucible, mix with 5 ml of H_2SO_4 (1 + 5) by means of a strong platinum wire, add 5 ml of pure hydrofluoric acid, and evaporate slowly on a sand bath under a good hood to sulphuric acid fumes begin to escape. Allow to cool, repeat the operation with fresh hydrofluoric acid until all gritty particles have disappeared, and expel most of the sulphuric acid; if necessary, add more H_2SO_4 (1 + 5) and evaporate again until all hydrofluoric acid has been expelled. Allow to cool, carefully dissolve the wet residue with water, warm the solution until clear, cool, transfer to a volumetric flask e.g., of 200 ml capacity, and adjust to volume with water.

Removal of the Ammonium hydroxide group

Transfer an aliquot of the sulphuric acid solution to a Pyrex flask, pass compressed air through a wash bottle containing ammonium hydroxide, and lead the vapour through a glass tube into the solution under test. Continue passing the gaseous ammonia until the solution is just neutralized as indicated by universal indicator paper. Centrifuge the Ammonium hydroxide group off, transfer the supernatant liquid to a volumetric flask, wash the precipitate with a few ml of water through which gaseous ammonia has been passed, and add the washings to the flask. If a further dilution is to be made, add sufficient acid just to acidify the solution. Otherwise add a fixed quantity of acid, e.g., H_2SO_4 (5 + 95), this being similar to that used in the standards.

Care should be taken to note the size of the aliquot taken both for the precipitation and for any further dilution used in the calculations.

Direct comparison of the sample solution with standards

When the transmission of solutions containing sodium and potassium is plotted against the respective concentrations up to ca. 50 p.p.m. a straight line is obtained. If this concentration is exceeded, the straight line becomes a curve due to self-absorption in the flame. Solutions should be diluted before measurement so that the concentrations lie within the range that will give a straight line relationship with concentration.

If the concentration of the sample solution is completely unknown, the solution should be compared directly with a standard solution, e.g., 50 p.p.m. and progressively diluted until it is below the standard.

Owing to the possibility of a drift in the electronic system of the instrument, comparison of the sample solution with the standards is made by placing them alternately in the flame.

Procedure

Turn the wave-length dial to the peak transmission for the constituent or insert the correct filter and then adjust the slitwidth so that the highest standard selected gives a reading high on the scale. Place the sample solution and standards alternately in the flame until the sample solution has been bracketed by standards. Repeat this procedure several times until constant readings are obtained. These readings are then plotted graphically as shown in Fig.18.

Standard addition

When this procedure is used it is preferable that background emission from the sample solution and standards is absent. If a background emission of any significance is present, this must be measured and subtracted as given below in the section on scanning.

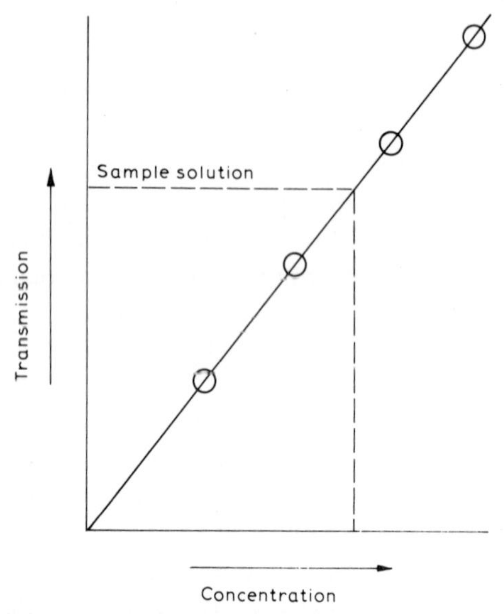

Fig.18. Transmission curve obtained by using standard solutions. Transmission of sample solution is compared with the curve to obtain concentration of alkali.

Procedure

An approximate value should already either be known or have been measured by comparing the emission of the sample solution directly with standard solutions as given above. Once this approximate value is known, either a single addition or a series of additions of standard solutions is made to the sample solution. To obtain the best relationship, the first addition should be such that the concentration in the sample solution is approximately doubled.

A suitable dilution of the sample solution is made, e.g., 10 ml aliquot diluted to 100 ml in a volumetric flask. Similar aliquots are placed in three other flasks and to these are added various volumes of a standard solution. The volumes of all flasks are then adjusted to 100 ml with water and placed in succession in the flame, with the wavelength dial turned to the peak wavelength. The slitwidth is

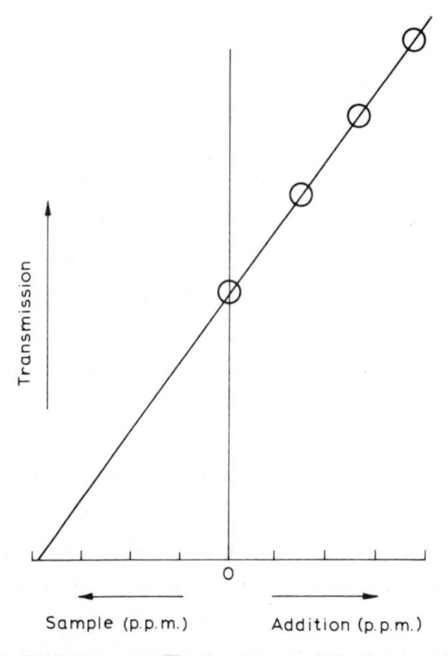

Fig.19. Standard addition curve. The intercept of the line projected onto the horizontal axis will give the concentration of alkali in the sample solution.

adjusted so that the highest reading i.e., sample solution plus largest standard addition, is accommodated by the scale. The four readings are then plotted graphically as shown in Fig.19.

The standard addition in p.p.m. is plotted along the base of the graph and the transmission readings of the solutions plotted vertically. The line joining the points is projected backwards to make an intersection on the base line from which the concentration in the sample solution is read.

The simple calculation shown below is used when only a single addition has been made: measure transmission of sample solution alone, transmission of sample solution plus standard and transmission of the blank (this is deducted from the other two readings). Concentration in sample solution in p.p.m. = concentration of standard in p.p.m. × reading of sample/ (reading of sample plus standard addition minus reading of sample).

Scanning

This procedure is appropriate when the sodium and potassium concentration is low, e.g. 0.1–1.0 p.p.m. in the solution.

The Ammonium hydroxide group may be separated by precipitation as given above. No other additions are made to the solution before measurement. The sample solution is usually undiluted and may have been concentrated by evaporation in a platinum basin to a smaller volume. Concentration does not necessarily improve the measurement since interfering constituents are also concentrated.

Procedure

An approximate value is first obtained by direct comparison with dilute standard solutions without any correction for background. Owing to the presence of background radiation at this level of concentration, the approximate value may be in excess of the true concentration by a factor of possibly two or three.

The instrument is set to the maximum sensitivity, e.g., using a Beckman D U the sensitivity knob is turned to the extreme left. The wave-length is turned to the appropriate peak for the constituent. The 1-p.p.m. standard solution is placed in the flame and the slit-width adjusted so that the transmission reading is ca. 100%. Standards of concentrations from 0.1–1.0 p.p.m. (at 0.1-p.p.m. intervals) and the sample solution are then placed successively in the flame and their transmissions recorded.

The wave-length is then adjusted to the off-peak setting i.e. 580 or 600 for sodium, 725 and 825 for potassium and the solutions again placed successively in the flame. The background transmission is then deducted from the former readings and a graph relating concentration of the standard solutions in p.p.m. with their respective peak heights is constructed. The peak height of the sample solution is then compared with this graph.

Deduction of the background emission, however, will not assist in the problem of depression of the emission by aluminium and magnesium. When aluminium has been removed by precipitation of the Ammonium hydroxide group, magnesium still remains in the solution. In the case of chondritic meteorite material (Easton and Lovering, 1964) a simple correction based on prepared mixtures of interfering elements was applied.

An alternative procedure is to apply the standard additions technique in conjunction with the scanning. The correction required to eliminate the interference caused by iron when measuring small quantities of potassium using a Perkin-Elmer model 146, has been given by Cooper et al. (1966).

CHAPTER 18

Water and Carbon Dioxide

CONSIDERATION OF METHODS

Uncombined water. Excluding natural glasses, the loss in weight of samples when heated for 2 h at 110°C is sufficient for the determination of uncombined water. During the preparation of sample water will usually be gained from the atmosphere, particularly during the crushing process. Fine powders i.e. <100 mesh will absorb a fraction of one per cent of water even when stored in a well stoppered bottle. If heavy liquids have been used in the separation of minerals, air drying should remove the residual liquid, but if any does remain it will be returned as combined water.

In the majority of analyses the uncombined water is only required in order to total the analysis.

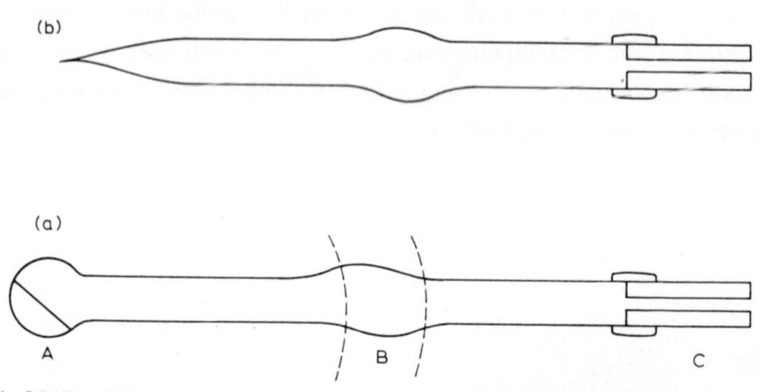

Fig.20. Penfield tube (*a*) for the determination of water. The sample is placed in the lower bulb *A*, and a band of wet paper is wrapped around *B* to assist condensation. A capillary tube *C* is used to restrict evaporation of water during the heating and weighing operations. The upper portion of the tube after being drawn off is illustrated by *b*.

Combined water. The determination of water and carbon dioxide was included in the loss on ignition in a number of the older analyses. Unfortunately decomposition of carbonates, oxidation of sulphides and of ferrous iron which is seldom complete, invalidates a determination by this method.

In the Penfield method (Penfield, 1894) for the determination of water, the weighed sample is heated in a hard glass tube with a capillary tube attached to the open end (Fig.20). The liberated water is condensed in the cooler portion of the tube. A small bulb is often blown half way along the tube to assist in collecting the droplets of water. The upper portion of the tube is drawn off and allowed to cool. This assembly is weighed, the capillary tube detached and the glass tube then dried in an oven set at 110°C. After cooling, the capillary tube is again attached and the assembly reweighed. The difference between the weights of the tube containing the condensed water and the tube after drying is the weight of water in the sample.

A modification of the Penfield method has been given (Shapiro and Brannock, 1962) in which the liberated water is absorbed on a weighed slip of filter paper. Another modification has been the absorption of the liberated water by a weighed desiccant, e.g., magnesium perchlorate (Harvey, 1939; Jeffrey and Wilson, 1960b). A simpler technique was given by Wilson (1962). In this case, the absorbent was contained in a weighed detachable absorption compartment connected by means of a ground glass joint to a silica ignition tube. The weighing compartment was sealed by a cap and plug while it was weighed.

It should be appreciated that none of the above methods allows the carbon dioxide to be determined.

Although the Penfield type methods may be both rapid and suitable for samples in which the water is released at a fairly low temperature i.e., 700°C, in some cases a higher temperature e.g. 1,100°C and a longer period of heating is required. Groves (1951, p.95) has described an apparatus where the sample, held in a platinum boat, is heated inside a silica tube by an electric furnace to a temperature in excess of 1,000°C. An inert gas carries the liberated water and carbon dioxide into weighed absorption tubes. The increase in weight of these tubes allows an accurate determination of these two constituents to be made.

Riley and Williams (1959b) have modified this apparatus to eliminate interference from sulphides if present in the sample and to reduce the size of the blank. Although the amount of apparatus required for this method is in excess of that required for the Penfield methods, once the apparatus has been set up, a large number of determinations may be made.

When minerals are present which only give up their water with difficulty, e.g., staurolite, a fusion may be necessary to release the water. A suitable flux weight for a 0.5 g sample suggested by Jeffrey and Wilson (1960b) was 2 g each of borax glass and anhydrous sodium tungstate. Using this flux staurolite was found to release its water after 30 min at 800°C.

Uncombined water

Procedure

Place a clean stoppered glass weighing bottle in an oven set at 110°C for 30 min; transfer it to a desiccator and allow it to cool for 30 min before weighing it empty. Place 2–5 g of sample in the weighing bottle and reweigh. A semi-micro analysis will use less, e.g., 50–100 mg. Place the bottle in the oven with the stopper slightly displaced for 2 h. At the end of this time, remove the weighing bottle from the oven, replace the stopper and allow it to cool in a desiccator for 30 min; then reweigh. The loss in weight is due to the uncombined water and is returned as a percentage.

This portion of the sample is usually reserved for determination of combined water and carbon dioxide.

Critical considerations

Special attention must be given to hydrated minerals and natural glasses. Determination of water "minus" has proved arbitrary in natural glasses (Drysdale et al., 1963). The value was found to be influenced by the following factors: mesh size of sample, duration of heating and temperature to which the sample was heated. This special class of material required the procedure given above to be modified as follows: (*1*) the determination should be made as soon as possible after the sample has been crushed; (*2*) if storage of the crushed sample is necessary then the container should be of minimum capacity; (*3*) the mesh size should be between 150–300 mesh and the mesh size used should be reported in the analysis; (*4*) the sample should be heated for 24 h at 110°C.

The use of a thermobalance is of assistance in investigations concerning natural glasses and hydrated minerals. The problem of clays has been dealt with by Mackenzie (1957). In this apparatus the weight-temperature relationship is recorded, permitting a detailed examination of the temperature at which the moisture is liberated. The temperature of the sample may either be held constant until the loss is negligible or increased at a constant rate. Fuller details of the thermogravimetric technique may be found in a work by Duval (1963).

Combined water

The general principle is given below, together with a detailed account of the various components of the absorption train.

General principle

The sample (0.1–5 g) is weighed into a platinum boat which is inserted into a silica tube contained in an electric furnace. The tube is flushed free of atmospheric air by means of a dry gas and the sample then moved into the heated zone of the furnace. The liberated water and carbon dioxide are flushed out of the tube into a previously weighed absorption tube train. After absorption is complete, the components of the train are reweighed and the increase in weight due to water and carbon dioxide noted.

Inert gas

The flushing medium must naturally be free of water and carbon dioxide; this is ensured by first passing the gas through a preliminary cleansing train (Fig.21). This train consists of tubes packed with soda lime anhydrous, magnesium perchlorate and finally phosphorus pentoxide; the first extracts carbon dioxide and the second and third extract any traces of water.

Water-free nitrogen ("dry") is a suitable gas for flushing the tube free of the initial atmosphere and transferring the liberated water and carbon dioxide vapour to the absorption train. Ordinary commercial nitrogen sometimes contains more water than the preliminary cleansing train can remove; it is for this reason that "dry" nitrogen is recommended.

The magnitude of the blank or completeness of flushing is check-

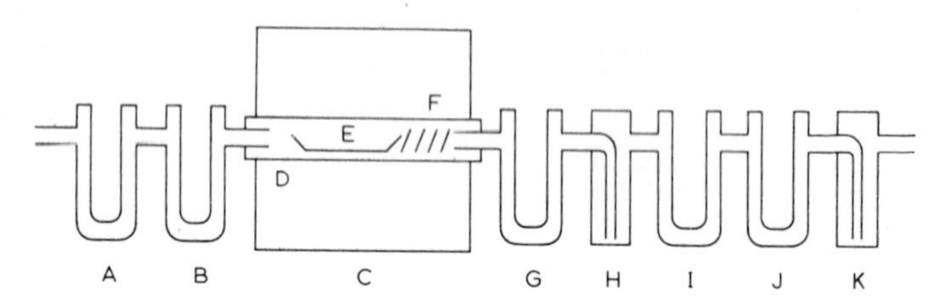

Fig.21. Absorption train for the determination of water and carbon dioxide. Components: *A* = soda lime; *B* = magnesium perchlorate in lefthand limb, phosphorus pentoxide in righthand limb; *C* = electric furnace capable of attaining 1,250°C; *D* = quartz or mullite tube; *E* = platinum boat containing sample; *F* = coil of copper wire; *G* = as in *B*; *H* = chromic acid bubbler; *I* = phosphorus pentoxide; *J* = soda lime in left hand limb, phosphorus pentoxide in right hand limb; *K* = sulphuric acid bubbler.

ed by passing the gas through the complete assembly for a known time, e.g., 30 min. The flow rate, 2–3 bubbles per second is measured by the final bubbler.

The blank is checked at least once a day and requires rechecking if the flow rate of the gas is altered.

Platinum boat

The size of the platinum boat should be ca. 4 X 1 cm for macro analysis and, if a choice is available, 3X0.5 cm for semi-micro analysis. Alternatively, an alumina boat may be lined with nickel foil and, if fluorine or sulphide is present, a covering of magnesium oxide previously ignited may be spread over the sample (Riley, 1958b). If the sample is only sintered, the cake should be tapped out gently. Under no circumstances should a sharp instrument be used to prise the cake from the platinum. If the sample has become fused to the boat during determination, a series of sodium carbonate fusions followed by immersion in a beaker of hydrochloric acid is the best method for its removal.

Occasionally a stain is left on the platinum, which may be removed by gently fusing a small quantity of potassium bisulphate in the boat. The fusion is removed when cold by placing the boat in a beaker containing hot H_2SO_4 (3 + 97). Care must be taken to ensure that all fusion mixtures are removed from the boat before the next determination.

The platinum boat may be placed inside a larger silica boat to facilitate insertion and removal from the tube.

Silica tube

The silica tube should be of the minimum diameter necessary to allow free movement of the platinum boat. It will be appreciated that if a larger than necessary size is used, then a longer time will be required to flush out both the initial atmosphere and the liberated gases.

The length should be sufficient to allow the platinum boat containing the sample to rest at <100°C while the tube is flushed out. The length should also be such that when the sample is in the heated zone, the exit end of the tube is sufficiently hot to avoid condensation of water before it reaches the absorption train. The optimum temperature for the exit end of the tube is between 40–60°C since a higher temperature will tend to decompose the rubber connection.

A spiral of copper foil or wire is placed inside the silica tube at the end of the hot zone to collect any sulphur compounds formed from decomposition of sulphides in the sample. Basic lead chromate and silver pumice powders have also been used for this purpose.

The closure of the silica tube may either be polytetrafluoroethylene or a good quality rubber bung but in each case the seal should be as near perfect as possible.

One method of moving the platinum boat into the hot zone is by means of a 1/16-inch stainless steel rod. A slight hook on one end assists in manoeuvering the boat into position. The rod passes through the closure at one end of the tube and, although a good seal is essential, it must also move freely. An objection to this method of moving the boat is the difficulty in ensuring that a good seal is obtained, for if a leakage does occur it will cause a low result.

An alternative method is to move the silica tube so that the boat is in the heated zone of the furnace.

Precleansing train

Since water is liberated during the absorption of carbon dioxide either by soda lime or soda lime asbestos, carbon dioxide is extracted first in *A* (Fig.21) and the water liberated by the reaction collected in the subsequent tube *B*. Tube *B* contains anhydrous magnesium perchlorate in the first arm and phosphorus pentoxide in the second.

The soda lime absorbent usually contains a small quantity of water. This is removed by passing dry carbon dioxide-free gas overnight through the absorbent before attaching it to the precleansing train.

Absorption train

The absorption train is made up of the following units:

G = contains anhydrous magnesium perchlorate in the left hand limb and phosphorus pentoxide in the right hand limb.

H = a bubbler containing saturated chromic acid in syrupy phosphoric acid.

I = a guard tube containing phosphorus pentoxide (this is not weighed).

K = a concentrated sulphuric acid bubbler which acts as a guard against the re-entry of moisture and indicates also the rate of flow of the gas through the absorption train.

For the above absorption tubes either a single limb or a double limb U tube may be used. The former is slightly easier to weigh on a semi-micro balance, while the latter reduces the number of weighings when two absorbents separated by a wad of glass wool are in a single tube.

The freshly prepared train is assembled and dry carbon dioxide-free gas is passed through the train overnight to remove moisture from the soda lime absorption tube. The phosphorus pentoxide may require renewing before a determination is made if the water content of the soda lime is high.

Chromic acid and phosphoric acid bubbler

A saturated solution of chromic acid in syrupy phosphoric acid is best prepared by placing acid an inch deep in the bottom of the bubbler and then adding a sufficient number of flakes of solid chromic acid to form a saturated solution.

Weighing of tubes

Once the absorption tubes have been filled with the appropriate reagent, the surface of the tubes should be wiped with paper tissue to remove adhering particles of reagent. A period of ca. 20 min should be allowed to elapse before weighing. The tubes should be stored overnight, or when not in use, in a desiccator and again allowed to stand for ca. 20 min on the bench so that equilibrium with the atmosphere is established before weighing. Provided the relative humidity of the air surrounding the absorption tubes is the same as that of the balance case and providing that cotton gloves are worn when handling them, satisfactory weighing are obtained.

If the practice of wiping the tubes with a chamois cloth is used, effects such as rate of reabsorption of moisture on the outside of the tube and static will make accurate weighing difficult.

Procedure

Weigh 0.1–5 g of sample into a platinum boat and insert the boat into the cool portion of the silica tube. Replace the closure and flush the silica tube free of atmosphere. Connect the preweighed absorption train and move the sample into the heated zone. After the period of heating, usually 1 h, disconnect the absorption train and reweigh the absorption tubes.

Determination of blank

The freshly prepared train is assembled and the dry carbon-dioxide-free gas passed through the absorption train, at the flow rate of 2–3 bubbles per sec, for 30 min. On reweighing the tubes no more than 1 mg should have been gained by the water absorption tube and probably slightly less by the carbon dioxide tube.

Interferences

When material is present which contains halogens (e.g. mica, amphibole), a flux consisting of lead oxide and magnesium oxide has been used to absorb fluorine and chlorine and to release the water quantitatively when heated with the sample to 800–900°C (Gillberg, 1964). The flux was prepared by heating equivalent quantities of lead and magnesium oxides in an electric furnace. The flux was added to the sample in the proportion of five parts of flux to one part of sample.

CHAPTER 19

Phosphorus

CONSIDERATION OF METHODS

Phosphorus does not occur native, because it readily combines with oxygen; it comprises about 0.12% of the lithosphere. In the form of phosphates it is found chiefly in metamorphic crystalline rocks, especially in metamorphosed limestone and in many metalliferous veins. Definite mineral phosphates of frequent occurence are e.g., apatite $3Ca_3(PO_4)_2.Ca(F,Cl)_2$, xenotime (yttrium phosphate), phosphorite (tricalcium phosphate), vivianite (normal hydrous ferrous phosphate), and wavellite (basic hydrous aluminium phosphate). Schreibersite (iron–nickel phosphide) occurs in meteoritic iron.

In silicate rocks and minerals, aluminium and iron being preponderant, all phosphorus in the sample caught in the weighed Mixed oxides (see Scheme I) can be determined after fusion with sodium carbonate, but is usually more conveniently determined in another portion of the sample, as is also the case with chromium, vanadium, rare earths metals, and zirconium, should these happen to be present in the Mixed oxides.

Using an independent portion of the sample it should be noted that: (*1*) some rocks and minerals containing phosphorus dissolve in mineral acids used single or in combination, whereas others are so insoluble that they must be broken up by fusing with fluxes; (*2*) in preparing solutions for the determination of phosphorus, the phosphorus must be quantitatively converted to orthophosphoric acid by appropriate oxidizing treatments such as digestion with nitric acid, or fusion with alkali; (*3*) fusions with alkalipyrosulphate, or boiling with concentrated sulphuric or perchloric acids, may cause a loss of phosphorus; (*4*) no losses of phosphorus occur when digestions with sulphuric acid are made at low temperatures to fume off acid, or when hydrochloric, nitric, or hydrofluoric acid solution are evaporated to dryness; (*5*) titanium and zirconium tend to retain the phosphorus as insoluble phosphate and, where these elements are present in appreciable amount, much of the phosphorus may be precipitated in the preliminary operations unless the insoluble residues are fused with, e.g., sodium carbonate, the phosphorus being extracted with water and added to the main solution.

The separation of phosphorus (together with aluminium, vanadium, and residual silicon) from iron and the other members of the Ammonium hydroxide group precipitated in the course of the General Procedure, can usually be made by treating an acid solution of the group with an excess of sodium hydroxide, or by fusing the Mixed oxides with alkali carbonate or peroxide, and leaching the melt with water.

Other methods commonly used for the separation of phosphorus from certain members of the Ammonium hydroxide group in silicate analysis are those based on: the removal of iron by extraction with ether; the removal of iron and chromium by electrolysis with a mercury cathode; the separation of iron, titanium, vanadium, and zirconium, by precipitation with cupferron; the separation of phosphorus by ion-exchange.

DECOMPOSITION OF SAMPLES CONTAINING PHOSPHORUS

In view of the foregoing it may be stated that for the determination of phosphorus in silicate rocks and minerals: (*a*) with samples which are for the most part quickly decomposed by mineral acids, an independent portion of the sample should preferably be subjected to a HF-HNO_3 attack; (*b*) with samples which resist to an acid attack, a separate portion of the sample must be broken up by fusing with Na_2CO_3, whereas the melt obtained should preferably be dissolved in nitric acid.

Attack by hydrofluoric and nitric acids
(see Scheme III)

Transfer 1 g of air-dry sample (–200 mesh to the linear inch) to a 100-ml flat-bottom platinum basin equipped with a platinum cover and add 25 ml of water, 25 ml of nitric acid, and about 20 ml of hydrofluoric acid (free from nonvolatile impurities) delivered directly from the polyethylene or ceresin bottle. Cover the basin, digest upon the steam bath for 30 min with occasional stirring with a platinum rod, remove the cover and evaporate the solution slowly on the bath under a good hood to dryness. Cool, carefully wash the inside surface of the basin with HNO_3 (1 + 1), bring the deposited salts into solution as much as possible, evaporate and re-evaporate with small quantities of nitric acid 3–4 times to dryness to decompose the fluorides and to expel fluorine. Cool, moisten the residue with 10 ml of nitric acid, add 100 ml of hot water and, while stirring, heat on the bath for 5–10 min. Allow any residue to settle somewhat, filter through a wet properly set No.42 Whatman or similar ashless filter paper into a graduated flask, transfer the residue to the paper by means of a jet of cool HNO_3 (1 + 99), wash paper and residue 5 times carefully and dropwise with the same dilute acid, detach the film of any insoluble matter adhering to the basin by means of a small piece of paper wrapped round a rubber-tipped glass rod and add the small piece of paper to the residue (Insoluble phosphates are not decomposed by evaporation with nitric acid and will be found in the residue). Reserve paper and residue (*1*) and also the combined filtrate and washings (*2*).

Ignite paper and residue (*1*) in a small platinum crucible under good oxidizing conditions at as low a temperature as possible, fuse with the minimum quantity of sodium carbonate, allow to cool, extract the melt with water, filter the extract through a small wet properly set No.42 Whatman ashless filter paper into a small beaker, rinse the crucible and wash paper and residue (*3*) a few times dropwise with sodium carbonate solution (1%), cover the beaker with a cover glass, acidify filtrate and washings (*4*) with nitric acid, and

SCHEME III
Dissolution of the sample by acids for the determination of phosphorus

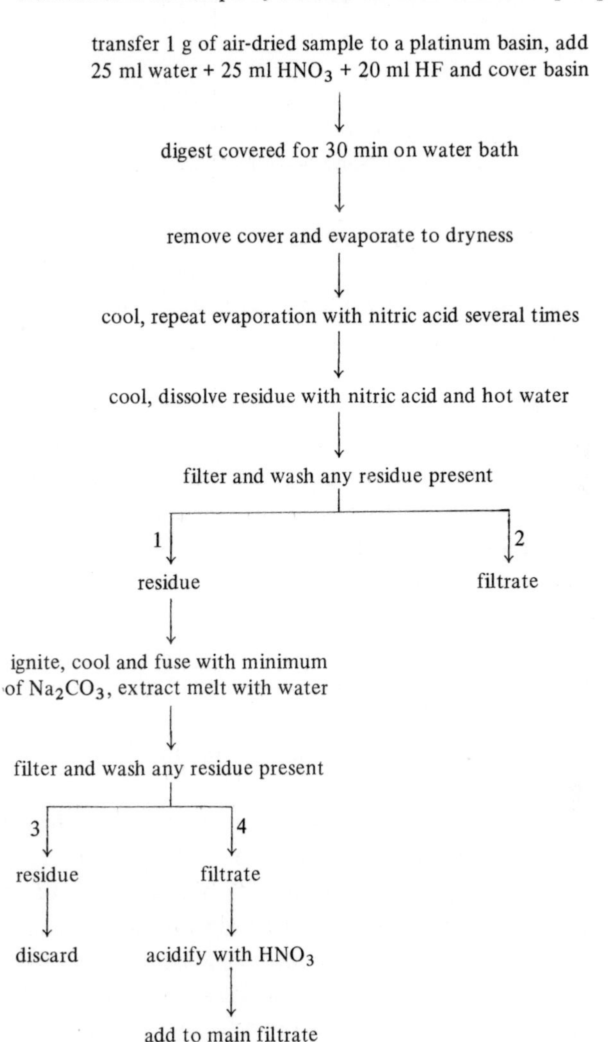

combine with the main filtrate and washings (*2*). Reserve the sample solution (i.e., the combined filtrates and washings *2,4*) for further treatment.

Fusion with sodium carbonate
(see Scheme IV)

Fuse 1 g of the air-dry sample (–150 mesh to the linear inch) with 4–6 g of pure anhydrous sodium carbonate as described in Chapter 5. Transfer the fusion cake from the crucible to a 300-ml platinum basin, add some water, cover the basin with a cover of platinum or pyrex glass, and add gradually under the cover 50 ml of HNO_3 (1 + 1). Place the basin on the steam bath and when disintegration is complete, remove the cover and wash any particles or solution adhering to the underside of the cover into the basin with a little water. Evaporate the dissolved carbonate melt and washings in the same basin and allow to stand on the bath for 2 h after approximate dryness, or longer, until the residue is free from fumes of nitric acid. Cover the basin, add cautiously to the dry and cool residue 10 ml of nitric acid, and after 2 min 100 ml of hot water. Remove the cover, wash the inside surface of the basin with a jet of hot water, heat on the bath for 5–10 min, and stir occasionally until the soluble salts are in solution. Allow any residue to settle somewhat, filter through a wet properly set No.42 Whatman or similar ashless filter paper into a graduated flask, transfer the residue to the paper by means of a jet of cool HNO_3 (1 + 99), wash paper and residue 10–12 times carefully and dropwise with the same dilute acid, detach the film of any insoluble matter adhering to the basin by means of a small piece of paper wrapped round a rubber-tipped glass rod and add the small piece of paper to the residue. Reserve paper and residue (*1*) and also the combined filtrate and washings (*2*).

Ignite paper and residue (*1*) in a small platinum crucible equipped with a platinum cover under good oxidizing conditions at as low a temperature as possible, carefully moisten the contents with HNO_3

SCHEME IV

Dissolution of the sample by fusion for the determination of phosphorus

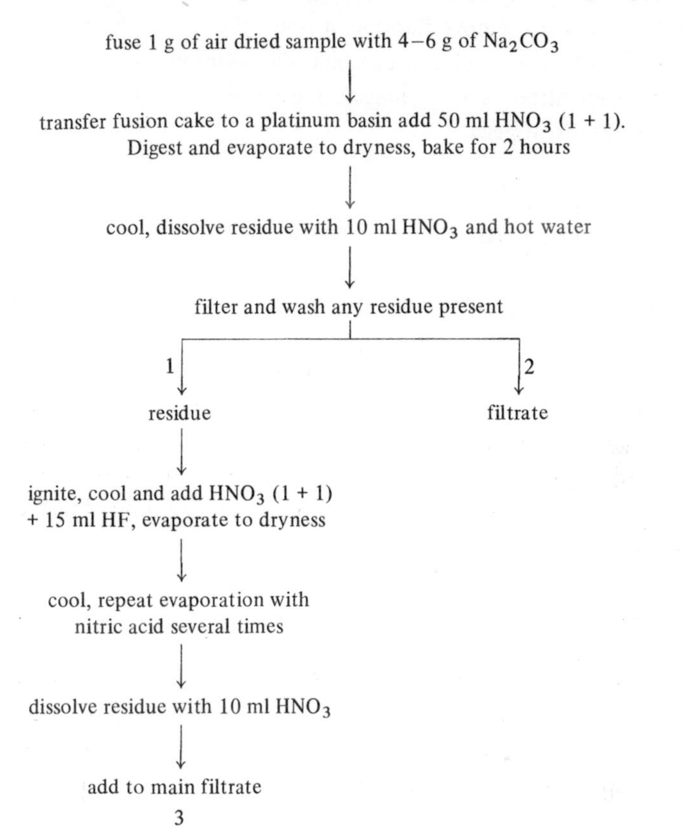

(1 + 1) by means of a small pipette inserted against the side of the crucible, and wash any particles or solution adhering to the underside of the cover into the crucible with a little of the same acid.

Next add 10–15 ml of hydrofluoric acid (free from nonvolatile impurities) delivered directly from the polyethylene or ceresin bottle, evaporate slowly upon the steam bath under a good hood to dryness to volatilize silicon tetrafluoride and the excess of hydrofluoric acid. Cool, wash the inside surface of the crucible with HNO_3

(1 + 1), bring the deposited salts into solution as much as possible, evaporate and re-evaporate with small quantities of nitric acid 3–4 times to dryness to remove fluorine, dissolve the residue by boiling with 10 ml of nitric acid and combine the solution and washings with the main filtrate and washings (*2*). Reserve the sample solution (*3*) for further treatment.

SEPARATION OF PHOSPHORUS BY THE ION-EXCHANGE METHOD

Principle of method

Cation are retained on the resin, phosphate-ion is eluted by means of dilute hydrochloric acid, the phosphate eluate evaporated to a convenient volume, and the amount of phosphorus determined by gravimetric, volumetric, colorimetric, or photometric methods. The procedure may be applied, e.g., to the sample solution obtained according to Scheme III or Scheme IV.

Preparation of the column

All reagents to be used must be phosphorous-free. Prepare a Dowex 50 X 2 (50–100 mesh to the linear inch) cation-exchange resin column (2 X 7 cm) by passing 100 ml of HCl (1 + 99) slowly through it; discard the liquid which has passed the column.

Procedure

Evaporate the sample solution, preferably obtained according to Scheme III, to dryness on the steam bath, convert the nitrates into chlorides by repeated evaporation with hydrochloric acid to dryness, cool, add 5 ml of hydrochloric acid and 45 ml of hot water, allow to stand on the bath for 5 min, cool, while stirring add NH_4OH (1 + 1) dropwise using thymol blue indicator paper to adjust the pH of the solution to 2.5–3.1, pour the solution slowly via a suitable fritted-glass filter through the prepared Dowex resin column, by which operation cations are absorbed on the resin, whereas the phosphate-ion should be washed down the column and removed from the resin by passing 150 ml of HCl (1 + 99) via the glass filter through the column, allowing to drain well between additions of the washing solution, and evaporate the eluate in a porcelain dish to a volume of 10–20 ml on the steam bath.

To remove any particles of resin that may have passed through the column, pour the concentrated eluate through a suitable fritted-glass filter into a smaller porcelain dish, wash the first dish and filter 6 times dropwise with HCl (1 + 99), evaporate the combined filtrate and washings to dryness on the bath, dissolve the residue with water containing a few drops of an appropriate acid and reserve the solution for the determination of phosphorus by one of the methods mentioned above.

(The metals held on the resin during the procedure may be released by washing the column with 100 ml of hydrochloric acid.)

SPECTROPHOTOMETRIC DETERMINATION OF PHOSPHORUS PENTOXIDE IN SILICATE ROCKS AND MINERALS

The two colorimetric methods commonly used are those involving the formation of phospho-vanado-molybdate, and the phospho salt of molybdenum blue. The former method is suitable for rocks and minerals also containing considerable amounts of phosphorus, whereas the latter being more sensitive is used in small-size samples when the material is scanty, or where phosphorus is present as a minor constituent. The latter method has also been applied to meteoric iron containing phosphorus in the form of phosphides (Moss et al., 1961).

A. By the molybdivanadophosphoric acid method

Principle of method

An excess of molybdate added to an acidified mixture of an ortho-phosphate and a vanadate forms a soluble yellow-coloured complex. Photometric measurement is made at approximately 430 mμ.

Concentration range

The recommended concentration range is from 0.04 to 0.3 mg of P_2O_5 in 25 ml of solution, using a cell depth of 1 or 4 cm.

Stability of colour

The time required for full development of the colour is dependent on the pH of the solution. Carried out as indicated, the yellow phosphovanadomolybdate colour develops within 5 min and is stable for at least 20 min.

Interfering elements

The interference of iron may be eliminated by using a similar size aliquot of the sample solution for the blank, adding 4 ml of nitric acid, and adjusting the volume with water to 25.0 ml. Chromium and nickel enhance the absorbancy; details of their interference and a general examination of variants in the method have been given (Lindley, 1961).

Special reagents

(*a*) Standard phosphate solution (1 ml ≡ 0.2 mg P_2O_5). – Dissolve 0.3834 g of KH_2PO_4 (dried at 110°C to constant weight) in a 1-l volumetric flask in ca. 250 ml of water, dilute with water to the mark, mix, and keep in a bottle of Pyrex glass.

(*b*) Standard phosphate solution (1 ml ≡ 0.02 mg P_2O_5). – Dilute one volume of phosphate solution (reagent *a*) tenfold with water in a volumetric flask, mix, and keep in a bottle of Pyrex glass.

(*c*) Mixed vanadate molybdate solution. – In a 1.5-l beaker marked at the 1-l level, dissolve 1.25 g of ammonium vanadate (NH_4VO_3) in 400 ml of cool HNO_3 (1 + 1). Prepare a second solution containing 50 g of ammonium molybdate [$(NH_4)_6Mo_7O_{24}, 4H_2O$] in 400 ml of water, and filter when solution is complete. While stirring, add the second solution to the first one, dilute the mixed solutions with water to the 1-l mark, mix, and keep in a dark-coloured, glass-stoppered bottle. (If on standing a precipitate forms, a fresh solution should be prepared.)

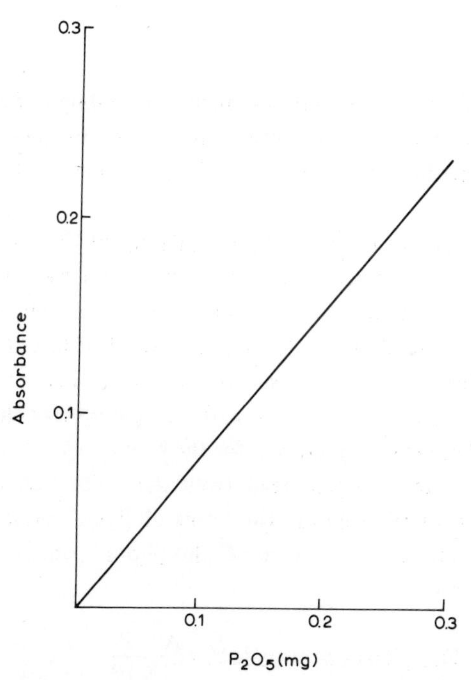

Fig. 22. Calibration curve for phosphorus (molybdivanadophosphoric acid method– using a 1-cm cell, 25 ml volume and a wave-length of 430 mμ.

Preparation of calibration curve

Transfer 2.0, 4.0, 6.0, 8.0, 10.0, and 15.0 ml of standard phosphate solution (reagent *b*) separately to six dry 50-ml glass-stoppered Erlenmeyer flasks, dilute with water to 15.0 ml and use 15.0 ml of water in an additional dry flask for the blank. To each flask add 10.0 ml of mixed vanadate molybdate (reagent *c*), stopper the flasks, mix, and let stand for 5 min.

Next transfer a suitable portion of the reference solution to an absorption cell having a 1- or 4-cm light path. Using a spectrophotometer measure the absorbancy at approximately 430 mμ, compensate or correct for the blank, and plot the values obtained against mg of P_2O_5 per 25 ml of solution (Fig.22).

Procedure

Dilute the clear sample solution obtained according to Scheme III or Scheme IV, freed from harmful cations, e.g., by means of the ion-exchange method, to a definite volume containing 10% v/v of nitric acid.

Transfer an aliquot ($\equiv$ 0.04–0.30 mg P_2O_5) of the diluted solution to a dry 50-ml glass-stoppered Erlenmeyer flask, dilute with water to 15.0 ml, and mix; use 15.0 ml of water in an additional dry flask for the blank. To both flasks add 10.0 ml of mixed vanadate molybdate solution (reagent *c*), stopper the flasks, mix, let stand for 5 min, and continue in accordance with "Preparation of the calibration curve". Compensate or correct for the blank.

Using the value obtained, read from the calibration curve the number of mg of phosphorus (in terms of P_2O_5) present in the aliquot. Calculate the percentage of phosphorus pentoxide of the sample as follows:

$$\text{Phosphorus pentoxide } \% = \frac{A - B}{C \times 10}$$

where:

A = mg of phosphorus pentoxide found in the aliquot used;
B = reagent blank correction in mg of phosphorus pentoxide;
C = grams of sample represented in the aliquot used.

B. By the molybdenumbluephosphoric acid method

Principle of method

Formation of the phospho compound of molybdenum blue under conditions of controlled acidity. Acidity range preferably 0.20–0.40 *N*. Photometric measurement is made at approximately 650 mμ.

Concentration range

The recommended concentration range is from 0.01 to 0.1 mg of P_2O_5 in 100 ml of solution, using a cell depth of 1 cm.

Stability of colour

The intensity and stability of the colour developed depend upon several factors including acid concentration, type of acid or salts present, and amount of reagent added. Carried out as indicated, the blue complex colour develops within ten minutes, and is stable for at least thirty minutes.

Interfering elements

Within the concentration range, aluminium, chromium, calcium, magnesium, manganese, nickel, and ferrous iron, do not interfere (Boltz, 1958). Aluminium in greater amounts retards development of the blue complex, ferric iron reduces the colour intensity.

Apparatus

Glassware. Owing to the sensitivity of the reaction, all glassware should be of the resistant variety. Riley (1958a) recommends that all glassware be filled with sulphuric acid and allowed to stand for several hours, next the flasks be emptied, be thoroughly rinsed with water, then filled with water and reserved for use in phosphorus determinations alone.

Special reagents

(*a*) Standard phosphate solution (1 ml ≡ 0.2 mg P_2O_5). – Prepare as in *A, a* (p. 220).

(*b*) Standard phosphate solution (1 ml ≡ 0.01 mg P_2O_5). – Dilute one volume of phosphate solution (reagent *a*) twentyfold with water in a volumetric flask, mix, and keep in bottle of Pyrex glass.

(*c*) Sodium sulphite solution (100 g Na_2SO_3/l). – Dissolve 10 g of anhyd. sodium sulphite in 100 ml of water. Prepare fresh as needed.

(*d*) Ammonium molybdate solution (20 g $(NH_4)_2MoO_4$/l). – Dissolve 4 g of ammonium molybdate in 80 ml of water and 120 ml of cold H_2SO_4 (1 + 1). Keep in dark-coloured bottle of Pyrex glass.

(*e*) Hydrazine sulphate solution (1.5 g $H_2N.NH_2.H_2SO_4$/l). – Dissolve 0.3 g of hydrazine sulphate in 200 ml of water. Prepare fresh as needed.

(*f*) Mixed molybdate hydrazine sulphate solution. – Immediately before use, dilute 50 ml of ammonium molybdate solution (reagent *d*) in a 200-ml volumetric flask with 110 ml of water. While swirling, add 20 ml of hydrazine sulphate solution (reagent *e*), dilute the mixed solutions with water to the mark, mix, and keep in dark-coloured bottle of Pyrex glass.

Preparation of calibration curve

Transfer 1.0, 2.0, 4.0, 6.0, 8.0, and 10.0 ml of standard phosphate solution (reagent *b*) separately to six 100-ml Erlenmeyer flasks, dilute with water to 10.0 ml and use 10.0 ml of water in an additional dry flask for the blank. To each flask add 10.0 ml of sodium sulphite solution (reagent *c*), mix, and boil gently for 10 min.

Next add to each flask 25.0 ml of mixed molybdate hydrazine sulphate solution (reagent *f*), mix, place the flasks on a steam bath for 40 min, cool to room temperature, transfer the solutions to seven 100-ml volumetric flasks, dilute with water to the mark, and mix.

Transfer a suitable portion of the reference solution to an absorption cell having a 1-cm light path. Using a spectrophotometer mea-

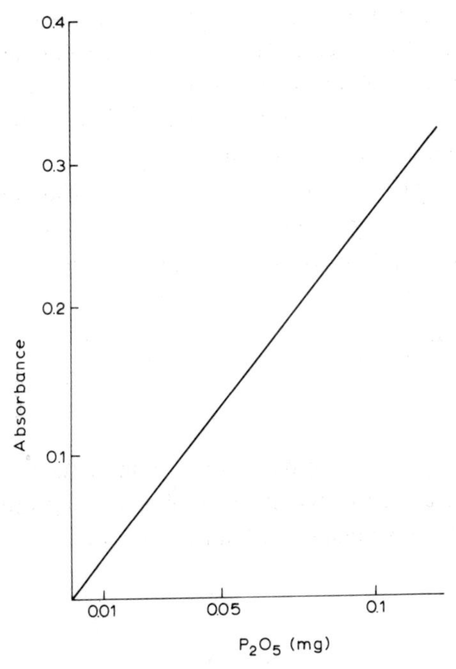

Fig.23. Calibration curve for phosphorus (molybdenumbluephosphoric acid method) using a 1-cm cell, 100 ml volume and a wave-length of 650 mμ.

sure the absorbancy at approximately 650 mμ, compensate or correct for the blank, and plot the values obtained against mg of P_2O_5 per 100 ml of solution (Fig.23).

Procedure

Dilute the clear sample solution obtained according to Scheme III or Schme IV, freed from harmful cations, e.g., by means of the ion-exchange method to a definite volume. (If much hydrochloric or nitric acid is present, remove most of it by evaporation!)

Transfer an aliquot (≡ 0.01–0.10 mg P_2O_5) of the diluted solution to a dry 100-ml Erlenmeyer flask, dilute with water to 10.0 ml, and mix; use 10.0 ml of water in an additional dry flask for the blank. To both flasks add 25.0 ml of mixed molybdate hydrazine sulphate solution (reagent *f*), mix, place the flasks on a steam bath for 40 min, cool to room temperature, transfer the solutions to two 100-ml volumetric flasks, dilute to the mark with water, mix, and continue in accordance with "Preparation of the calibration curve". Compensate or correct for the blank.

Using the value obtained, read from the calibration curve the number of mg of phosphorus (in terms of P_2O_5) present in the aliquot. Calculate the percentage of phosphorus pentoxide of the sample as follows:

$$\text{Phosphorus pentoxide } \% = \frac{A - B}{C \times 10}$$

where:

A = mg of phosphorus pentoxide found in the aliquot used;
B = reagent blank correction in mg of phosphorus pentoxide;
C = grams of sample represented in the aliquot used.

CHAPTER 20

Total Sulphur

CONSIDERATION OF METHODS

Sulphur comprises over 0.05% of the lithosphere. It is found native (e.g., in the regions of active and extinct volcanos as a volcanic sublimate formed by reactions between sulphur dioxide and hydrogen sulphide), in the oxidized condition as sulphate radical (e.g., in minerals of the barite group, and in gypsum and alunite), or as sulphate group (e.g., as a component of the orthosilicates haüynite and nosalite containing 14% SO_3), and widely distributed in the form of sulphide (pyrites, blendes and glances, in the orthosilicate lazurite containing 17% S, and in minerals found in the more basic rocks). Because of the fact that in silicate rocks sulphur occurs for the most part in the sulphide condition, it is recommended to report the sulphur therein as S.

The two fundamental methods for the determination of sulphur in silicates and silicate rocks are the gravimetric method in which sulphur is converted by an oxidizing agent into a salt of sulphuric acid and weighed as barium sulphate, and the evolution method in which sulphur is evolved as hydrogen sulphide, fixed in ammoniacal solution as a metallic sulphide and after the addition of acid, oxidized by a standard solution of iodate or iodine. The gravimetric method is used in the most exact analyses, the volumetric as a rapid routine method of limited application. As the determinations according to the General procedure do not leave residues or solutions that can be conveniently used in the determination of sulphur, this element is usually determined in a separate portion of the sample by methods that are subject to interference by only a few elements.

A cation-exchange separation of sulphate from the other major constituents may be required for certain chondrite material, e.g., for triolite. The procedure is similar to that given for phosphate (pp.219–220) where the cations are adsorbed on the resin while the sulphate is eluted. The eluate is evaporated to dryness on a steam bath and the residue dissolved in water containing a few drops of hydrochloric acid.

A turbidimetric determination may be of use for small amounts of soluble sulphate but this method does involve a number of critical factors described by Patterson (1957). These factors include: crystalline form of the precipitate, particle size, distribution of light-scattering particles, presence of certain foreign ions (e.g., lithium-ions), temperature and ageing of the suspension before measurement, and the excess of barium chloride in the solution.

Precipitation of the soluble sulphate as benzidine sulphate is an alternative method suitable for small quantities. The precipitate is centrifuged off and dissolved in HCl (1.5 + 98.5), diazotized and coupled with N-1(1-naphtyl) ethylenediamine dihydrochloride. The resulting purple colour is measured spectrophotometrically. Vogel (1958, p.669) notes that the presence of chlorides and excess phosphates cause incomplete precipitation of benzidine sulphate.

Gravimetric determination of Total sulphur in silicate rocks and minerals, containing no more than 2.5 per cent Total S

The chief requirements for the accurate determination of Total sulphur in silicate rocks and minerals as barium sulphate are: (*1*) to prevent air-oxidation of sulphide material during the preparation of the sample; (*2*) to convert all the sulphur to SO_4^{2-} ion; (*3*) to obtain a pure barium sulphate; (*4*) to reduce the solubility of the barium sulphate.

As regards the first, fine grinding may cause slight losses of sulphur as the dioxide and formation of sulphate; therefore, crushing of the sample (without grinding!) should be carried so far as to furnish the coarsest powder that can be decomposed completely by the method of attack. Fussions with Na_2CO_3 and a little KNO_3 (the latter to an amount little greater than is needed to oxidize fully the small amount of sulphides present) decompose the well-crushed sample with no loss of sulphur as hydrogen sulphide or elemental sulphur, and only slightly attack the platinum crucible. In a mixture of elements in solution, a precipitate of pure barium sulphate is seldom obtained because the sulphate ion does not react excusively with the barium ion. Moreover, the precipitate is apt to contain small amounts of occluded barium chloride, ferric and alkali ions, possibly as $(BaCl)_2SO_4$ and $BaFe_2(SO_4)_4$ causing a plus error, and alkali ions causing a minus error, in the determination. A minus error is also caused by the solubility of barium sulphate; the solubility increases with the concentration of acid which should not exceed that needed to prevent hydrolyses of the salts that may be present. A filter paper of close texture must be used for filtration, and the amount of washing solution that is used should be held down to the minimum that is needed. No silica will be separated with the barium sulphate when the alkaline filtrate is neutralized with hydrochloric acid and cooled, diluted to about 200 ml with water, acidified in the cold with 5 ml of HCl (1 + 1), and the sulphur is precipitated at boiling heat by barium chloride in excess; suspecting contamination by silica, the barium sulphate precipitate can be treated with HF-H_2SO_4 before the barium sulphate is weighed. When the Total sulphur content of the sample does not exceed 2.5%, and the precipitation as sulphate is made in the absence of iron according to the procedure given below, no special precautions have usually to be made and no special purification of the barium sulphate precipitate will be necessary. The minus error caused by the solubility of the precipitate can almost be corrected by means of a blank determination (see Procedure).

Special reagents

(*a*) Sodium sulphate solution (1 ml ≡ 0.1 mg S). – Transfer 0.4428 g of anhydrous Na_2SO_4 to a 1-l volumetric flask, dissolve and dilute to the mark with water.

(*b*) Methyl orange indicator solution (1 g/l). – Dissolve 0.1 g of methyl orange in 100 ml of water; filter if necessary.

(*c*) Barium chloride solution (100 g $BaCl_2 2H_2O$/l). – Dissolve 100 g of crystallized barium chloride in water and dilute to 1 l.

Procedure

Mix thoroughly, using a platinum rod, 2 g of the air-dry sample crushed to 100–150 mesh to the linear inch, with about 10 g of purest anhydrous sodium carbonate and a little potassium nitrate in a platinum crucible (KNO_3 to an amount little greater than is needed to oxidize all the oxidizable components of the sample). Protecting the sample by means of an asbestos shield from direct contact with the products of combustion, and with occasional stirring with a platinum rod, fuse carefully over a Bunsen flame until the melt is quiet, and then heat for some time over an inclined blast. Transfer the fusion cake from the crucible to a 200-ml Pyrex beaker, add 100 ml of cool water and 2 drops of ethyl alcohol (to reduce and precipitate any permanganate present), stir until complete disintegration of the cake, let settle, filter the supernatant liquid through a wet properly set 11-cm No.40 Whatman or similar ashless paper into a 300-ml Pyrex beaker (marked at the 200-ml level), and wash crucible, paper and residue, by decantation with a dilute sodium carbonate solution (5 g Na_2CO_3/l) in small portions. (At this stage the quantitative determination of sulphur – as well as that of chromium – in the filtrate, may be combined with that of barium, zirconium, and rare earths in the sodium precipitate, if desired.) The filtrate may be yellow from sodium chromate if chromium is present in appreciable amounts.

Dilute the filtrate and washings to the mark with water, add 10.0 ml of sodium sulphate solution (reagent *a*) and a few drops of methyl orange indicator (reagent *b*), neutralize with HCl (1 + 1) while stirring with a rubber-tipped glass rod, cool, add 5 ml of HCl (1 + 1), cover the beaker with a watch glass, and heat to boiling until action ceases. Remove the beaker from the source of heat, remove and rinse the cover with water and, if clear, add 10 ml of barium chloride solution (reagent *c*) dropwise while stirring vigorously. Cover the beaker, let stand at the side of the steam bath for 2 h, and then overnight at room temperature.

Decant the solution through a fresh properly set 9 cm close-texture ashless paper into a 300-ml beaker; if clear, discard the filtrate. Next transfer funnel and paper to a 200-ml flask, wash the precipitate 3 times by decantation with hot water in small portions, transfer the precipitate quantitatively to the paper, and wash with hot water dropwise until almost free from chlorides; the precipitate clinging to the beaker be detached by means of a small piece of ashless paper wrapped round the rubber-tipped glass rod and added to the precipitate. (Complete removal of chloride ions may be tested by collecting a few drops from the last washing in a test tube containing 2 ml of a 0.1 *N* solution of silver nitrate and 1 drop of 4 *N* nitric acid; a white curdy precipitate of AgCl is entirely soluble in ammonia.) The volume of the combined washings should be measured in aid of the blank determination.

Transfer the paper and precipitate to a weighed platinum crucible, dry on the steam bath, char the paper without inflaming, next heat at 900°C, let cool over sulphuric acid in a desiccator and weigh as $BaSO_4$. (If contamination by silica is suspected, the ignited precipitate can be treated with some drops of hydrofluoric acid and one drop of sulphuric acid before the barium sulphate is weighed.) Reweigh until constant weight is obtained, and correct for the minus error due to the solubility of $BaSO_4$, determined by means of the blank.

Blank. Make a blank determination, following the same procedure and using the same amounts of all reagents and washings, including the 10.0 ml of sodium sulphate solution (reagent *a*).

Calculate the percentage of Total sulphur of the sample as follows:

$$\text{Total sulphur } \% = \frac{(A - B) \times 0.1374}{C} \times 100$$

where:

A = grams of $BaSO_4$ from the sample;

B = grams of $BaSO_4$ from the blank;

C = grams of sample used.

CHAPTER 21

Formulas of Minerals

DISCUSSION

The main use of atomic formula calculation as far as the analyst is concerned, is for checking the correctness of the chemical analysis of a mineral. Within certain limits, errors in the chemical analysis will give an imperfect balance in the calculations.

Although the exact placing of certain constituents is still debatable, sufficient is known for the major constituents to be allocated in the formula (Hess, 1949). Often Ti, Mn, P and Cr are present in minerals in low concentrations such that the molecular proportion is <0.001. Although these may be ignored for the purpose of the calculations, the accuracy of each determination should still be high since these will give a good indication of trends of other trace elements present in the mineral.

The method used for calculating the atomic formula is given in detail, together with several worked examples. Two illustrations are given of how an error may be present in an analysis without this being indicated by the formula. From these it will be seen that calculations are not a complete test for an accurate analysis but are helpful in indicating large errors.

CALCULATIONS

The calculations may be made in the following manner, as set out in Table IV. The analytical percentages are tabulated in column I and are returned where possible to two decimal places. The percentage of each constituent is divided by its molecular weight shown in Table V

TABLE IV
Stages in the calculation of atomic formula

Column	*1*	*2*	*3*	*4*	*5*
	%	*mol. ratio*	*cation ratio*	*anion ratio*	*cation ratio to x oxygens*
Oxide					
MO		1	1	1	values from
M_2O		1	2	1	column 3 ×
M_2O_3		1	2	3	factor
M_2O_5		1	2	5	
MO_2		1	1	2	
				Sum	

x no of oxygens/sum = factor

TABLE V
Molecular weights of the major constituents of silicate material

Oxide	*Mol. weight*	*Oxide*	*Mol. weight*
SiO_2	60.09	MgO	40.32
TiO_2	79.90	Cr_2O_3	152.02
Al_2O_3	101.96	P_2O_5	141.95
Fe_2O_3	159.70	Na_2O	61.99
FeO	71.84	K_2O	94.19
MnO	70.93	OH	17.00
CaO	56.08	F	19.00

and tabulated in column 2. Where the quotient is less than 0.001, that fact is entered and no further calculation made on that particular constituent.

The third column is the cation ratio in which the molecular ratio from column 2 is adjusted for the number of metal atoms in the oxide e.g. doubled for aluminium, ferric iron, potassium, sodium and phosphorus. For the oxide in which only one metal atom is present the figure is identical to that in column 2.

The fourth column is the anion ratio in which the molecular ratio

TABLE VI

Atomic formula groupings of some common silicate minerals

Mineral	*No. of oxygen atoms*	*Group*			
		1	*2*	*3*	*4*
nthophyllite	24	Si, Al (8)	Al, Ti, Fe^{3+}, Mg, Fe^{2+}, Mn, Ca, Na, K (7)	OH, F (1–4)	
ugite	6	Si, Al (2)	Al, Ti, Fe^{3+}, Cr^{3+}, Mg, Fe^{2+}, Mn, Ca, Na, K (2)		
iotite	24	Si, Al, Ti (8)	Al, Ti, Fe^{3+}, Fe^{2+}, Mn, Mg (6)	Ca, Na, K (2)	OH, F (1–5)
ustamite	18	Si, Al (6)	Mg, Ti, Fe^{2+}, Mn, Ca (6)		
hromite	32	Si, Al, Cr, Fe^{3+}, Ti (16)	Mg, Fe^{2+}, Mn, Ca (8)		
arnet	24	Si, Al (6)	Al, Fe^{3+}, Ti (4)	Mg, Fe^{2+}, Mn, Ca, Na, K (6)	
lauconite	24	Si, Al (8)	Al, Ti, Fe^{3+}, Cr, Fe^{2+}, (4)	Ca, Na, K (1.2–2.0)	
ornblende common)	24	Si, Al (8)	Al, Ti, Fe^{3+}, Mg, Fe^{2+}, Mn (5)	Ca, Na, K (2)	OH, F (0–2)
llite	24	Si, Al (8)	Al, Ti, Fe^{3+}, Fe^{2+}, Mg, Mn, Ca (4)	Na, K (1–2)	OH (4–5)
adeite	6	Si (2)	Al, Fe^{3+}, Ti (1)	Mg, Fe^{2+}, Mn, Ca, Na, K (1)	
Kaolinite	18	Si (4)	Al, Ti, Fe^{3+}, Fe^{2+}, Mg (4)	Ca, Na, K (< 1)	OH (8)
Magnetite	32	Si, Al, Cr, Fe^{3+}, Ti (16)	Mg, Fe^{2+}, Mn, Ca (8)		
Muscovite	24	Si, Al (8)	Al, Ti, Cr, Fe^{3+}, Fe^{2+}, Mn, Mg (4)	Ca, Na, K (2)	

TABLE VI
(continued)

Mineral	No. of oxygen atoms	*Group*			
		1	*2*	*3*	*4*
Olivine	4	Si (1)	Al, Ti, Fe^{3+}, Fe^{2+}, Mn, Ca (2)		
Omphacite	6	Si, Al (2)	Al, Ti, Fe^{3+}, Mg, Fe^{2+}, Mn (1)	Ca, Na, K (1)	
Orthopyroxene	6	Si, Al, Ti, Fe^{3+}, Ca (2)	Al, Ti, Fe^{3+}, Cr, Mg, Fe^{2+}, Mn, Ca, Na, K (2)		
Pectalite	18	Si, Al, Fe^{3+}, (6)	Mg, Fe^{2+}, Mn, Ca, (4)	Na, K, OH (2)	
Phlogopite	24	Si, Al, Ti (8)	Al, Ti, Fe^{3+}, Fe^{2+}, Mn, Mg (6)	Ca, Na, K (2)	
Pigeonite	6	Si, Al, Ti (2)	Fe^{3+}, Mg, Fe^{2+}, Mn, Ca, Na, K (2)		
Rhodonite	18	Si, Al (6)	Al, Fe^{3+}, Mg, Fe^{2+}, Mn, Ca (6)		
Sphene	20	Si, Al (4)	Al, Fe^{3+}, Mg, Ti, Fe^{2+}, (4)	Mn, Na, Ca, K (4)	
Spinel	32	Si, Al, Cr, Fe^{3+}, Ti (16)	Mg, Fe^{2+}, Mn, Ca (8)		
Smectite	24	Si, Al (8)	Al, Ti, Fe^{3+}, Fe^{2+}, Mg (4)	Ca, Na, K (0.7–1.0)	
Talc	24	Si, Al (8)	Al, Fe^{3+}, Fe^{2+}, Mn, Mg (6)	Ca, Na, K (<1)	OH (
Tourmaline	31	Si, Al (13–15)	Al, Fe^{3+}, Mg, Ti, Fe^{2+}, Mn, Ca, Na, K (4–5)	OH, F (3–4)	

The groupings given in this table are based on modern structural formulae (Deer et al., 1963); summation of group shown in brackets.

is adjusted in accordance with the number of oxygen atoms present in the oxide, e.g. the molecular ratio is doubled for silica and titanium; trebled for chromium and ferric iron; quintupled for phosphorus. The figure is identical to that in column 2 for the oxide with only one atom of oxygen.

A summation of column 4 is made and divided into the number of oxygen atoms present in the general formula of the mineral, a selection of which are shown in Table VI.

The number present in column 3, i.e., cation ratio is then multiplied by the resultant of the division and tabulated in column 5.

When the table has been completed, various groupings of the atomic ratios are made as indicated in Table VI. The calcium/magnesium/iron ratio may be calculated for correlation with optical data. Other ratios such as the percentage of aluminium occupying the tetrahedral position with silica may also be calculated from the atomic formula. Examples of the calculation of the atomic formula applied to an olivine, a clinopyroxene and a garnet are shown in

TABLE VII

Atomic formula calculation applied to an olivine (X_2ZO_4) analysis

Oxide	*%*	*Mol. ratio*	*Cation ratio*	*Anion ratio*	*Cation ratio to 4 oxygens*
SiO_2	39.75	0.662	0.662	1.324	0.971
TiO_2	0.21	0.003	0.003	0.006	0.009
Al_2O_3	1.48	0.015	0.030	0.045	0.066
FeO	9.63	0.134	0.134	0.134	0.196
MnO	0.13	0.002	0.002	0.002	0.003
MgO	48.51	1.203	1.203	1.203	1.764
CaO	0.67	0.012	0.012	0.012	0.017
Na_2O	0.006	<0.001			
K_2O	0.01	<0.001			
P_2O_5	0.07	<0.001			
Cr_2O_3	0.01	<0.001			
Total	100.48				

*) Z = Si + Al to make 1; X = 2.026

Tables VII, VIII and X. Two examples illustrating the limitations of the calculation of the atomic formula in exposing errors are shown in Table IX. The letters W, X and Y are used to represent groupings of elements within the mineral formula; while the letter "p" is used to express the ratio between the "W" and the "XY" group of a clinopyroxene. These figures (Table IX) show that the presence of this error does not disturb the summation of WXY sufficiently to indicate its presence in the analysis. However, it will be noticed that the value of "p" in example 1 does not balance to the same degree as that in the correct analysis given in Table VIII. In example 2, since

TABLE VIII

Atomic formula calculation applied to a clinopyroxene $[(W)_{1-p}XY_{1+p}Z_2O_6]$ analysis

Oxide	%	*Mol. ratio*	*Cation ratio*	*Anion ratio*	*Cation ratio to 6 oxygens*
SiO_2	47.44	0.789	0.789	1.578	1.750
TiO_2	1.09	0.011	0.011	0.022	0.024
Al_2O_3	10.76	0.106	0.212	0.318	0.470
Fe_2O_3	3.76	0.024	0.048	0.072	0.106
FeO	5.15	0.072	0.072	0.072	0.160
MnO	0.12	0.002	0.002	0.002	0.004
MgO	11.20	0.278	0.278	0.278	0.617
CaO	18.13	0.323	0.323	0.323	0.717
Na_2O	2.11	0.034	0.068	0.034	0.151
K_2O	0.03	0.001	0.001	0.001	0.002
P_2O_5	0.15	0.001	0.002	0.005	0.004
Cr_2O_3	0.02	<0.001			
Total	99.97				

W = Ca 0.717	X = Mg 0.617	Y = Al 0.220	Z = Si 1.750
Na 0.151	Fe^{2+} 0.160	Fe^{3+} 0.106	Al 0.250
K 0.002	Mn 0.004	Ti 0.024	
0.870	0.781	0.341	2.000
WXY = 1.992	p = 0.125 (± 0.005)		
$Ca_{44.8}$	$Mg_{38.6}$	$Fe_{16.6}$	

TABLE IX

Examples of alteration to the atomic formula values when certain errors are assumed to be present in the analysis of the clinopyroxene analysis (see in Table VIII)

Example 1. CaO increased by 1%, MgO decreased by 1%

Z		W		X		Y	
Z = Si	1.755	W = Ca	0.758	X = Mg	0.562	Y = Al	0.217
Al	0.245	Na	0.151	Fe^{2+}	0.160	Fe^{3+}	0.106
		K	0.002	Mn	0.004	Ti	0.024
Total	2.000		0.911		0.726		0.347
WXY = 1.984		p = 0.081 (± 0.08)					

Example 2. SiO_2 increased by 1%

Z		W		X		Y	
Z = Si	1.766	W = Ca	0.707	X = Mg	0.609	Y = Al	0.230
Al	0.234	Na	0.149	Fe^{2+}	0.158	Fe^{3+}	0.105
		K	0.002	Mn	0.004	Ti	0.024
Total	2.000		0.858		0.771		0.359
WXY = 1.988		p = 0.136 (± 0.006)					

the constituents of W and X are unchanged in their relative proportions, a reasonable balance is still obtained for the value of "p".

The atomic formula calculated from a garnet analysis may be used for representing the analysis in the form of end-members of the series. These are listed in Table XI together with the controlling elements. This form of presentation of a garnet analysis is useful for geochemical studies.

The calculation is made by placing the major atomic ratios on the line immediately under the respective element, Table XII. From this set of figures, the atomic ratio of the controlling element of the first end-member i.e. almandite is subtracted. The atomic ratios of the other elements also present in this end-member, in their correct proportions, given in Table XI are then also subtracted. For example, in almandite where Fe^{2+} is the controlling element, the ratio

TABLE X

Atomic formula calculation applied to a garnet ($W_3X_2Z_3O_{12}$)

Oxide	*%*	*Mol. ratio*	*Cation ratio*	*Anion ratio*	*Cation ratio to 24 oxygens*
SiO_2	40.33	0.672	0.672	1.344	6.08
TiO_2	0.26	0.003	0.003	0.006	0.03
Al_2O_3	20.74	0.204	0.408	0.612	3.69
Fe_2O_3	2.28	0.014	0.028	0.042	0.25
FeO	17.66	0.246	0.246	0.246	2.22
MnO	0.43	0.006	0.006	0.006	0.05
MgO	10.89	0.271	0.271	0.271	2.45
CaO	7.08	0.126	0.126	0.126	1.14
Na_2O	0.08	0.001	0.002	0.001	0.02
K_2O	0.05	0.0005	0.001	0.0005	0.01
P_2O_5	0.04				
Total	99.84				

W		X		Z	
W = Fe^{2+}	2.22	X = Al	3.69	Z = Si	6.08
Mn	0.05	Ti	0.03		
Mg	2.45	Fe^{3+}	0.25		
Ca	1.14				
Na	0.02				
K	0.01				
	5.89		3.97		6.08

TABLE XI

End-members of the garnet series

End-member	*Formula*	*Controlling element*
Almandite	$Fe^{2}_{6}Al_4Si_6O_{24}$	Fe^{2+}
Andradite	$Ca_6(Fe^3, Ti)_4Si_6O_{24}$	$(Fe^{3+} + Ti)$
Spessarite	$Mn_6Al_4Si_6O_{24}$	Mn
Uvarovite	$Ca_6Cr_4Si_6O_{24}$	Cr
Grossular	$Ca_6Al_4Si_6O_{24}$	Ca
Pyrope	$Mg_6Al_4Si_6O_{24}$	Mg

TABLE XII

Representation of a garnet analysis (Table XI) in the form of end-members

Elements	Fe^{2+}	(Fe^{3+}, Ti)	*Mn*	*Cr*	*Ca*	*Mg*	*Si*	*Al*		
Atomic ratio to 24 oxygens	2.22	0.28	0.05	0.01	1.14	2.45	6.08	3.69		
Almandite	2.22						2.22	1.48	5.92	37.9%
		0.28	0.05	0.01	1.14	2.45	3.86	2.21		
Andradite		0.28			0.42		0.42		1.12	7.2%
			0.05	0.01	0.72	2.45	3.44	2.21		
Spessarite			0.05				0.05	0.03	0.13	0.8%
				0.01	0.72	2.45	3.39	2.18		
Uvarovite				0.01	0.01		0.01		0.03	0.1%
					0.71	2.45	3.38	2.21		
Grossular					0.71		0.71	0.48	1.90	12.2%
						2.45	2.67	1.73		
Pyrope						2.45	2.45	1.63	6.53	41.8%
					Sum of end-members =				15.63	100%

subtracted for Al is two thirds of the value of Fe^{2+} while the ratio subtracted for Si is equal to that of the Fe^{2+}.

In the case of andradite where $(Fe^{3+} + Ti)_4$ is the controlling element, the ratio subtracted for Ca and Si is one and a half times that of $(Fe^{3+} + Ti)$.

The ratios of each end-member are subtracted in turn until all of the ratios are accounted for the calculation. The summation of the atomic ratios of each end-member is placed at the right hand side of the table. These in turn are totalled and the percentage of each end-member calculated.

POSSIBLE POINTS FOR CHECKING A CHONDRITE METEORITIC ANALYSIS

Certain ratios and values appear to dominate in chondritic meteorites and these may prove to be of some assistance in assessing the correctness of the analysis. It must be stressed however, that although they appear to be the general rule, in the light of present knowledge exceptions may occur.

Since the use of flamephotometric techniques for determination of potassium and sodium, more accurate values for these two elements have become available. These show that sodium (Na) is greater than potassium (K) by a ratio of from 8–12 in chondritic material, excluding several of the carbonaceous chondrites which have a lower ratio (Edwards and Urey, 1955). A comparison of the ratio between these two elements is preferred as this is independent of inhomogeneity in the distribution of the metallic and silicate phases.

Improved titanium values (Greenland and Lovering, 1965) have shown that in chondritic meteorites, excluding the enstatites, a fairly constant value of between 500 and 600 p.p.m. Ti is present. In the enstatitic class the values are more variable being from 300–700 p.p.m. as Ti.

Ahrens (1964) has shown that the majority of chondrites have a constant ratio of silicon to magnesium within the various classes. These values were found to be 1.43 for carbonaceous, 1.92 for enstatites and 1.60 for ordinary chondrites.

CHAPTER 22

Determination of the Specific Gravity

ROCKS

Two types of samples are usually submitted for the determination of the specific gravity, i.e., bulk rock and mineral grains. A knowledge of the specific gravity of the latter is often helpful in the selection of grains for analysis.

Bulk rock samples are usually of sufficient volume to allow a simple displacement method to be used. If pure and free from impurities (internal and external), and not porous, the sample is suspended by a thin wire and weighed first in air and then again while immersed in a liquid of known density. A correction is made for the weight of the wire used for the suspension. If water is used, this should be boiled and the flask stoppered while cooling to ensure that it is free from dissolved air. Tables relating the density of water with temperature are to be found in a number of standard chemistry text books, e.g., Vogel (1958). Alternatively, organic liquids of high density such as ethyltetrabromide may be used provided that the specific gravity of the liquid is either already known or measured.

Where the sample is porous, a portion not required for analysis may be coated with a water impervious material such as shellac before the determination is made.

The formula used for calculating the specific gravity is: weight of sample in air $\times$ density of liquid used for suspension/weight of sample in air minus weight of the sample in the liquid.

MINERAL GRAINS

The specific gravity of mineral grains may be determined by the flotation method in which the specific gravity of the liquid is adjusted until the grains neither rise nor fall. This method is rather time-consuming since adjustment of the specific gravity of the liquid must be made, either by the addition of a miscible liquid of differing specific gravity, or by an alteration in the temperature of the liquid. If the former method is used, it is still necessary to measure the specific gravity of the mixture. Where a change in temperature is used to alter the specific gravity of the suspending liquid, a previously prepared curve relating temperature to specific gravity of the liquid may be used.

The most rapid method of specific gravity measurement is that using a torsion balance with the attachments shown in Fig.24. These may either be bought or made by spot welding of thin platinum

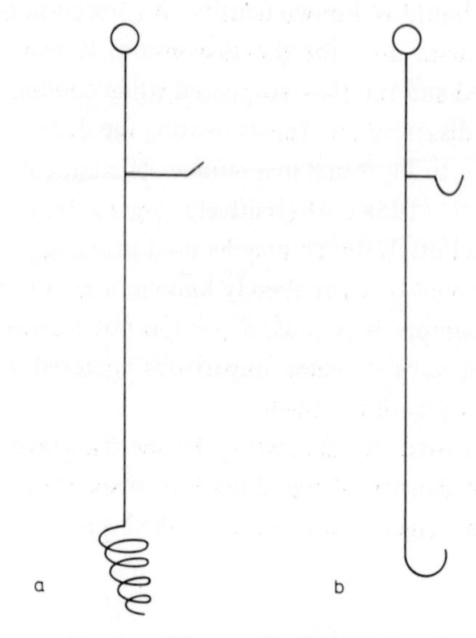

Fig.24. Attachments for use with a torsion balance for the determination of gravity of mineral grains.

wire, foil and gauze. A suitable balance should have a range of 100 mg. The dial adjustment is normally 25 mg, calibrated in 0.1 mg divisions with a vernier reading 0.01 mg. Detachable riders of 25 and 50 mg are used to complete the range. Where the attachment weight is less than 25 mg, grains weighing up to 75 mg may be used.

Mineral grains which do not fall within this weight range may be broken either in a small steel or agate mortar depending upon the size and hardness of the grains. Care should be taken that the grain is not pulverized in the process. A needle-pointed pair of tweezers is useful for handling grains and transfer should always be made over a sheet of white paper to avoid loss of the sample on the open bench.

Toluene is often used in preference to water since it is sufficiently volatile to allow rapid drying of the sample, has a fairly low surface tension and is obtainable in a pure form of known specific gravity. It is good practice when weighing the attachments and sample, to obtain the point of balance each time while the attachment is moving in a certain direction. Errors due to a thin film of toluene on the attachments are thereby compensated. Immersion of the spiral or basket should be to a depth of 3–5 mm. The attachment *a* shown in Fig.24 is suitable for larger mineral grains which are of a sufficient size to be placed on the plate. A worked example for this attachment is illustrated in Table XIII.

TABLE XIII

A worked example of a specific gravity determination using a torsion balance

Temperature of toluene	= 20°C
Specific gravity of toluene	= 0.863
Weight of assembly plus sample on plate	= 71.61 mg
Weight of assembly alone	= 25.19 mg
Difference = weight of sample in air	= 46.42 mg
Weight of assembly plus sample in spiral	= 61.18 mg
Weight of assembly alone	= 25.19 mg
Difference = weight of sample in toluene	= 35.99 mg
Specific gravity	= 3.84

Specific gravity = weight of sample in air times S.G. of toluene at 20°C/weight of sample in air minus weight of sample in toluene.

Where the mineral grains are either too small and numerous to place on the plate then the basket assembly *b* as shown in Fig.24 is used. The assembly is first weighed empty with the basket on the upper hook while the lower hook is immersed in the toluene to a depth of 3–5 mm. This is the weight of the assembly in air. The basket is then transferred with the aid of the tweezers to the lower hook and the assembly reweighed. This is the weight of the assembly in toluene. After allowing the toluene to evaporate, a small quantity of the sample is placed in the basket; the basket, plus sample placed on the upper hook and reweighed. The basket is then carefully transferred to the lower hook and weighed again. The specific gravity of the sample is usually returned to two decimal places, the third figure being used to correct the second place.

Where the mineral grains are too fine to be retained by the mesh of the basket, a bucket of similar size may be constructed from thin platinum foil. When using this attachment it is preferable to wet the mineral grains with toluene from a glass dropper before immersing the bucket in the toluene, to ensure complete wetting of the grains.

CHAPTER 23

Notes on the Precision and Accuracy of Results Obtained in Silicate-Rock Analysis and the Limit of Allowable Error

The traditional practice of inspecting the summation of a rock analysis is a necessary test of its correctness, but a summation near to 100.0% is no assurance of its accuracy because the analysis is subject to many errors. As principal causes of these may be mentioned: the non-homogeneity of the rock; contamination of the laboratory sample by iron, agate, etc., from crushing, sieving and sampling apparatus; chemical changes by fine grinding of oxidizable materials or fixation of water or carbon dioxide; difficulties in the way of mixing, riffling, and quatering to obtain a representative test sample; the use of inaccurate weights and inexactly calibrated volumetric equipment as well as unavoidable errors inherent in the methods of analysis to be used. Weighed portions taken from the same powdered test sample by a single analyst will never be chemically identical because of the difference in specific gravity of its components, of which the measure of precision of selecting generally deteriorates with decrease in the concentration of the component to be determined.

Regarding the developments in instrumental technique, a number of so-called standard methods can be displaced by modern procedures, e.g., involving a chromatographic separation and a spectrophotometric determination or flame photometry and absorption spectroscopy. But for the reasons given below, and for the convenience and benefit of a tyro, those standard analytical procedures have been maintained which have been demonstrated to be reliable, to have convenient practical application and at the same time give reproducible results in the hands of less experienced laboratory workers. In relation to this, it should be remembered that for the determination of the major constituents of rocks, gravimetric or volumetric methods are mostly preferable, since they are generally more accurate when large quantities are involved (see p. 2). Also, that for micro and semimicro quantities of the minor constituents photometric methods (with their advantages of speed and relative high precision) may be used, although even under the most favourable circumstances no higher accuracy can be obtained than about one per cent of the amount present in the photometric determination.

In view of the sources of accidental and probable errors mentioned above, and in the absence of knowledge of the actual composition of the rocks, no very precise estimate can be made of the errors to be expected; consequently it is difficult to prescribe the accuracy with which an analyst should determine a constituent if a certain method is applied to a certain material. Therefore, because for lack of practical data, and regarding the variations in duplicate determinations of the same constituent, and pending the final limits to be laid down by further interlaboratory work and improvement of analytical procedures, it should be recommended for the present to make it a rule to hold to the following limits of variability, calculated in terms of the whole rock: for constituents which amount to 25% and over, 0.2–0.3%; for constituents which amount to 10–25%, 0.1–0.2%; for constituents to 1–10%, 0.05–0.1%; and for those which amount to some tenths of one per cent only, 0.02% at the most. These limits do not hold for the analysis of silicate minerals; with this, higher grades of precision and accuracy can be obtained because of the more homogeneous character of the test sample.

Besides the allowable differences and limits recommended, the summation of all constituents should range from 100.0–100.3%, because in a complete silicate-rock analysis with close approach to accuracy, there is always a tendency towards too high results. This phenomenon is caused by the following: reagents in the method of chemical analysis of rocks, although carefully prepared, are seldom absolutely pure and their impurities plus those derived through their attack on the vessels used, will be weighed in whole or in part with the constituents of the rocks; retention of alkalies after fusion by incomplete washing of precipitates; improper ignition of precipitates, and failure to protect the ignited residues from moisture before and during weighing; also the dust entering the test samples during their dissolution and evaporation of their solutions, will give rise to high values of the final products to be weighed. Therefore, failure to attain 100.0% in the greater part of a series of analyses, should be an indication that in the course of the examination something has been overlooked.

In evaluating the precision of any method (as a measure of the duplicability of the method in the hands of a single analyst), reliance may be placed on so-called "standard methods of analysis" designed and adopted by competent chemists, and being intended for general use.

The accuracy of any method (as a measure of degree of correctness of the method) can only be judged by the use of standard samples duplicating the material under test, because of demonstrating certain errors being inherent in the method (as e.g., those caused by imperfect separation or solubility of the precipitates) affecting both sample and standard sample to the same extent.

Analyzed by a sufficient number of methods and research chemists, the certificate of composition of a standard sample is unquestionable reliable, and the use of standard samples, resembling as closely as possible in chemical nature the material under test, is of great help in testing new procedures or checking methods to be applied.

References

Ahrens, L. H., 1964. Si–Mg fractionation in chondrites. *Geochim. Cosmochim. Acta*, 28: 411–423.

Andersson, L. H., 1962. Studies in the determination of silica. 7. Some experiments with the gravimetric silica determination. *Arkiv Kemi,* 19: 249–256.

Belcher, C. B., 1963. Sodium peroxide as a flux in refractory and mineral analysis. *Talanta*, 10: 75–81.

Berg, R., 1927, Koplexe Metallverbindungen des Oxychinolins. *J. Prakt. Chem.,* 115: 180.

Berry, H. and Rudowski, R., 1965. The preparation of chondritic meteorites for chemical analysis. *Geochim. Cosmochim. Acta*, 29: 1367–1369.

Betz, J. B. and Noll, C. A., 1950. Total hardness determination by direct colorimetric titration. *J. Am. Water Works Assoc.*, 42: 49–56.

Bisque, R. E., 1961. Analysis of carbonate rocks for calcium, magnesium, iron and aluminium with EDTA. *J. Sediment. Petrol.*, 31: 113–122.

Bradfield, E. G., 1962. A study of some factors which affect the adsorption of titan yellow on magnesium hydroxide. *Anal. Chim. Acta*, 27: 262–271.

Brannock, W. W. and Berthold, S. M., 1949. The determination of sodium and potassium in silicate rocks by flamephometer. *U. S. Geol. Surv., Bull.*, 992.

Burke, R. W. and Yoe, J. H., 1962. Simultaneous spectrophotometric determination of cobalt and nickel with 2,3-quinoxalinedithiol. *Anal. Chem.*, 34: 1378–1382.

Chirnside, R. S., 1960. Review of silicate analysis. *Glass Technol.*, 43: 5 T.

Cooper, J., 1963. The flamephotometric determination of potassium in geological materials used for potassium argon dating. *Geochim. Cosmochim. Acta*, 27: 525–546.

Cooper, J. A., Martin, I. D. and Vernon, M. J., 1966. Evaluation of rubidium and iron bias in flame photometric potassium determination for K–Ar dating. *Geochim. Cosmochim. Acta*, 30: 197–205.

Davis, C. E. S., 1961. A rapid method of analysing cements and rock products. *Australia Commonwealth Sci. Ind. Res. Organ. Chem. Res. Lab., Tech. Paper*, 3.

Dean, J. A., 1960. *Flamephotometry*. McGraw-Hill, New York, N.Y., 354 pp.

Deer, W. A., Howie, R. A. and Zussman, J., 1963. *Rock Forming Minerals*. Longmans, London, 1125 pp.

Den Boef, G., De Jong, W. J., Krijn, G. C. and Poppe, H., 1960. Spectrophotometric determination of chromium (3) with EDTA. *Anal. Chim. Acta,* 23: 557–564.

Diehl, H., 1940. *The Applications of the Dioximes to Analytical Chemistry.* G. Frederick Smith Chem. Co., Ohio.

Dinnin, J. I., 1959. Rapid analysis of chromite and chrome ore. *U.S. Geol. Surv., Bull.*, 1084–B.

Drysdale, D. J., Lacy, E. D. and Tarney, J., 1963. "Water minus" (H_2O^-) in natural glasses. *Analyst*, 88: 131–133.

Duval, C., 1963, *Inorganic Thermogravimetric Analysis.* Elsevier, Amsterdam, 772 pp.

Easton, A. J., 1964. The determination of chromium in the presence of manganese in rocks and minerals. *Anal. Chim. Acta,* 31: 189–191.

Easton, A. J. and Greenland, L., 1963. The determination of titanium in meteoritic material. *Anal. Chim. Acta,* 29: 52–55.

Easton, A. J. and Lovering, J. F., 1963. The analysis of chondritic meteorites. *Geochim. Cosmochim. Acta,* 27: 753–767.

Easton, A. J. and Lovering, J. F., 1964. Determination of small quantities of potassium and sodium in stony meteorites material, rocks and minerals. *Anal. Chim. Acta,* 30: 543–548.

Edwards, G. and Urey, H. C., 1955. Determination of alkali metals in meteorites by a distillation process. *Geochim. Cosmochim. Acta,* 7: 154–168.

Feigl, F., 1946. *Spot Tests. 1. Inorganic Application.* Elsevier, Amsterdam.

Flaschka, A., 1953. Direct volumetric determination of divalent manganese with EDTA in presence of other metals. *Chemist-Analyst*, 42: 56–58.

Gillberg, M., 1964. Halogens and hydroxyl contents of mica and amphiboles in Swedish granitic rocks. *Geochim. Cosmochim. Acta*, 28: 495–516.

Goto, H., Kakita, Y. and Namiki, M., 1957. A new spectrophotometric determination of titanium with sodium alizarinsulfonate. *J. Chem. Soc. Japan, Pure Chem. Sect.*, 78: 373–376.

Goto, K., Komatsu, T. and Furukawa, T., 1962. Rapid colorimetric determination of manganese in waters containing iron. *Anal. Chim. Acta,* 27: 331–334.

Greenland, L. and Lovering, J. F., 1965. Minor and trace element abundances in chondritic meteorites. *Geochim. Cosmochim. Acta,* 29: 821–858.

Groves, A. W., 1951. *Silicate Analysis.* Allen and Unwin, London.

Habasky, M. G., 1961. The quantitative determination of metallic iron in the presence of iron oxides in treated ores and slags. *Anal. Chem.*, 33: 586–588.

Harvey, C. O., 1939. 2. Simple method for determination of water in silicates. *Bull. Geol. Surv. Gr. Brit.*, 1: 8–12.

Heisig, G. B., 1928. Volumetric determination of ferrous iron by means of potassium iodate. *Am. Chem. Soc.*, 50: 1687–1691.

Hess, H. H., 1949. Chemical composition and optical properties of common clinopyroxenes, 1. *Am. Mineralogist*, 34: 621–666.

Hey, M. H., 1941. The determination of ferrous iron in resistant silicates. *Mineral. Mag.*, 26: 116–118.

Hillebrand, W. F. and Lundell, G. E. F., 1953. *Applied Inorganic Analysis.* Wiley, New York, N.Y., p.639.

Hoops, G. H., 1964. The nature of the insoluble residues remaining after the $HF-H_2SO_4$ acid decomposition (solution B) of rocks. *Geochim. Cosmochim. Acta*, 28: 405–406.

Ingamells, C. O., 1960. A new method for "ferrous iron and excess oxygen" in rocks, minerals and oxides. *Talanta*, 4: 268–273.

Jeffery, P. C. and Wilson, A. D., 1960a. A combined gravimetric and photometric procedure for determining silica in silicate rocks and minerals. *Analyst*, 85: 478–485.

Jeffery, P. C. and Wilson, A. D., 1960b. Closed circulation systems for determining water, carbon dioxide and total carbon in silicate rocks and minerals. *Analyst*, 85: 749–755.

Kameswara Rao, V., Sundar, D. S. and Satri, M. N., 1965. Photometric determination of chromium using EDTA. *Chemist-Analyst*, 54: 86.

Kinnunen, J. and Wennerstrand, B., 1957. Some further applications of xylenol orange as an indicator in the EDTA titration. *Chemist-Analyst*, 46: 92–93.

Kitson, R. E. and Mellon, M. G., 1944. Colorimetric determination of phosphorus as molybdivanadophosphoric acid. *Ind. Eng. Chem. Anal. Ed.*, 16: 379–383.

Klassova, N. S. and Leonova, L. L., 1964. Absorptiometric determination of titanium with 3,6-dichlorochromotropic acid in small samples of minerals and rocks. *Zh. Analit. Khim.*, 19: 131.

Kolthoff, I. M. and Sandell, E. R., 1947. *Textbook of Quantitative Analysis.* Revised ed. 1947. Macmillan, London.

Kolthoff, I. M. and Stenger, V. A., 1947. *Volumetric Analysis, II.* ed. 1947. Interscience Publishers, New York.

Konig, H., 1964. Über die chemische Analyse von Chondriten. *Geochim. Cosmochim. Acta*, 28: 1697–1703.

Korbl, J. and Pribil, R., 1956. Xylenol orange: new indicator for the EDTA titration. *Chemist-Analyst*, 45: 102–103.

Lewis, L. L., Nardozzi, M. J. and Melnick, L. M., 1961. Rapid chemical determination of aluminium, calcium and magnesium in raw material, sinters and slags. *Anal. Chem.*, 33: 1351–1355.

Libermann, A., 1955. The determination of small amounts of nickel in copper ores and concentrates containing iron and cobalt. *Analyst*, 80: 595–598.

Lindley, G., 1961. The absorption determination of phosphorus in irons and steels. *Anal. Chim. Acta*, 25: 334–342.

Lovering, J. F., Nichiporuk, W., Chodos, A. and Brown, H., 1957. The distribution of gallium, germanium, cobalt, chromium and copper in iron and stony-iron meteorites in relation to nickel content and structure. *Geochim. Cosmochim. Acta*, 11: 263–278.

Mackenzie, R. C., 1957. *The Differential Thermal Investigation of Clays.* Mineral. Soc., London, 456 pp.

McLaughlin, R. J. W. and Biskupski, V. S., 1965. The rapid determination of silica in rocks and minerals. *Anal. Chim. Acta*, 32: 165–169.
23–39.

Milner, G. W. C. and Woodhead, J. L., 1954. The volumetric determination of aluminium in non ferrous alloys. *Analyst*, 79: 363–367.

Moss, A. A., Hey, M. H. and Bothwell, D. I., 1961. Methods for the chemical analysis of meteorites. 1. Siderites. *Mineral. Mag.*, 32: 802–816.

Moss, A.A., Hey, M.H., Elliott, C.J. and Easton, A.J., 1967. Methods for the chemical analysis of meteorites, 2. The major and some minor constituents of chondrites. *Mineral Mag.*, 36: 101–119.

Murmann, E., 1910. Über die Trennung von Kalk und Magnesia. *Z. Anal. Chem.*, 49: 688.

Nicholls, G. D., 1960. Techniques in sedimentary geochemistry. 2. Determination of the ferrous iron contents of carbonaceous shales. *J. Sediment Petrol.*, 30: 603–612.

Nichols, P. N. R., 1960. The photometric determination of titanium with tiron. *Analyst*, 85: 452–453.

Nordling, W. D., 1962. Photometric determination of manganese in high chromium steels. *Chemist-Analyst*, 51: 14–15.

Patterson, G. D. Jr., 1957. Sulphur. In: D. F. Boltz (Editor), *Colorimetric Determination of Non Metals.* Interscience, New York, N.Y., p.261.

Patton, J. and Reeder, W., 1956. New indicator for titration of calcium with ethylenedinitrilotetraacetate. *Anal. Chem.*, 28: 1026–1028.

Penfield, S. L., 1894. On some methods for the determination of water. *Am. J. Sci.*, 48: 30–37.

Penner, E. M. and Inman, W. R., 1961. Flame photometric methods used in the mineral sciences division, mines branch, Ottawa. *Can. Dep. Mines Tech. Surv., Mines branch, Tech. Bull T. B*, 24.

Poeder, B. C., den Boef, G. and Franswa, C. E. M., 1962. Selective spectrophotometric determination of iron (III) with EDTA. *Anal. Chim. Acta*, 27: 339–344.

Prior, G. T., 1913. The meteoric stones of Baroti, Punjab, India and Wittekrantz, South Africa. *Mineral. Mag.*, 17: 22–32.

Rafter, T. A., 1950. Na_2O_2 decomposition of minerals in platinum vessels. *Analyst*, 75: 485–492.

Rigg, T., and Wagenbauer, H. A. 1961. Spectrophotometric determination of titanium in silicate rocks. *Anal. Chem.*, 33: 1347–1349.

Riley, J. P., 1958a. The rapid analysis of silicate rocks and minerals. *Anal. Chem. Acta*, 19: 413–428.

Riley, J. P., 1958b. Simultaneous determination of water and carbon dioxide in rocks and minerals. *Analyst*, 83: 42–49.

Riley, J. P. and Williams, H. P., 1959a. The micro-analysis of silicate and carbonate minerals. 1. Determination of ferrous iron. *Mikrochim. Acta*, 4: 516–524.

Riley, J.P. and Williams, H.P., 1959b. The micro-analysis of silicate and carbonate minerals. 2. Determination of water and carbon dioxide. *Mikrochim. Acta*, 4: 525–535.

Rowland, E. O., 1963. A simple sample divider. *Mineral. Mag.*, 33: 524.

Rowledge, H. P., 1934. A new method for the determination of ferrous iron in refractory silicates. *J. Roy. Soc. W. Australia,* 2: 165–199.

Sandell, E. B., 1959. *Colorimetric Determination of Traces of Metals.* Interscience, New York, N.Y., p.872.

Selmer-Olsen, A. R., 1962. Spectrophotometric determination of chromium with 1,2-diaminocyclohexanetetracetic acid. *Anal. Chim. Acta,* 26: 482–486.

Shapiro, L. and Brannock, W. W., 1962. Rapid analysis of silicate, carbonate and phosphate rocks. *U. S. Geol. Surv., Bull.*, 1144–A.

Shaw, D. M., 1961. Element distribution laws in geochemistry. *Geochim. Cosmochim. Acta,* 23: 116–134.

Smith, G. R. and Bonnick, W. M., 1958. The Fe (II) and Co (II) chelation complexes of 2,6-bis(4-ethyl-2-pyridyl)-4-phenyl-pyridine, *Anal. Chim. Acta*, 18: 269–271.

Snell, F.D. and Snell, C.J., 1959. *Colorimetric Methods of Analysis Including Photometric Methods. 2A.* Van Nostrand, London.

Spencer, L. J. and Hey, M. H., 1932. Hoba (S. W. Africa) the largest known meteorite. *Mineral. Mag.*, 23: 1–18.

Stevens, R. E. and Niles, W. W., 1960. Chemical analyses of the granite and dibase. 1. Second report on a cooperative investigation of the composition of two silicate rocks. *U. S. Geol. Surv., Bull,* 1113: 3–43.

Stott, V., 1928. *Volumetric Glassware.* Witherby, London.

Swift, E.H., 1924, A new method for the separation of gallium from other elements. *Am. Chem. Soc.*, 46: 2378.

Thiel, A. und Peter, O., 1935. Sulphosalicylic Acid, Grundlagen und Anwendungen der Absolutcolorimetrie. 12. Die absolutcolorimetrische Bestimmung des Eisens. *Zh. Analit. Khim,* 103: 161–166.

Treadwell, F. P. and Hall, W. T., 1935–7. *Analytical Chemistry,* 2 vols. Chapman and Hall, London.

Tuma, J., 1962. Optimum conditions for the colorimetric microdetermination of silicon. *Mikrochim. Acta,* 3: 513–523.

Van Klooster, H. S., 1921. Nitroso-R-salt a new reagent for the detection of cobalt. *J. Am. Chem. Soc.,* 43: 746–749.

Van Loon, J. C., 1965. Titrimetric determination of the iron (2) oxide content of silicates using potassium iodate. *Talanta,* 12: 599–603.

Van der Walt, C. F. J. and Van der Merwe, 1938. Colorimetric determination of chromium in plant ash, soil, water and rocks. *Analyst,* 63: 809–811.

Van Wesemael, J. Ch., 1961. The determination of magnesium with titan yellow. *Anal. Chim. Acta,* 25: 238–247.

Vincent, E. A., 1960. Analysis by gravimetric and volumetric methods, flamephotometry, colorimetry and related techniques. In: A. A. Smales and L. R. Wager (Editors), *Methods in Geochemistry.* Interscience, New York, N.Y. p.49.

Vogel, A. I., 1954. *Macro and Semi-micro Qualitative Inorganic Analysis.* Longmans, London, 680 pp.

Vogel, A. I., 1961. *Quantitative Inorganic Analysis.* Longmans, London, 1248 pp.

Von Baudisch, O., 1909. Quantitative Trennung mit "Cupferron". *Chem. Ztg.*, 33: 1298.

Wahl, W., 1950. The statement of chemical analyses on stony meteorites and the interpretation of the analyses in terms of minerals. *Mineral. Mag.*, 19: 416–426.

Washington, H.S., 1919–30. *Chemical Analysis of Rocks.* Chapman and Hall, London.

Willard, H. H. and Greathouse, L. H., 1917. The colorimetric determination of manganese by oxidation with periodate. *J. Am. Chem. Soc.*, 39: 2366–2377.

Williams, C. H., 1960. The suppression of calcium interference in the flame photometric determination of exchangeable sodium in soils. *Anal. Chim. Acta*, 23: 183–185.

Wilson, A. D., 1955. A new method for the determination of ferrous iron in rocks and minerals. *Bull. Geol. Surv. Gt. Brit.*, 9: 56–58.

Wilson, A. D., 1960. The micro determination of ferrous iron in silicate minerals by a volumetric and a colorimetric method. *Analyst*, 85: 823–827.

Wilson, A. D., 1962. Determination of total water in rocks by a simple diffusion method. *Analyst*, 87: 598–600.

Wilson, A. D., 1964. The sampling of silicate rock powders for chemical analysis. *Analyst*, 89: 18–30.

Wilson, A. D. and Sergeant, G. A., 1963. The colorimetric determination of aluminium in minerals by pyrocatechol violet. *Analyst*, 88: 109–112.

Wilson, A. D., Sergeant, G. A. and Lionnel, L. J., 1963. The determination of total sulphur in silicate rocks by wet oxidation. *Analyst*, 88: 138–140.

Yoe, J. H. and Armstrong, A. R., 1947. Colorimetric determination of titanium with disodium-1,2-dihydroxybenzene-3,5-disulphonate. *Anal. Chem.*, 19: 100–102.

Index